Tales of Physicists and Mathematicians

Tidal Theories and Mathematical

Semyon Grigorevich Gindikin

Tales of Physicists and Mathematicians

Translated by Alan Shuchat

With 30 Illustrations

Birkhäuser
Boston · Basel

Semyon Grigorevich Gindikin
A.N. Belozersky Laboratory of Molecular
   Biology and Bioorganic Chemistry
Moscow State University
Moscow 119899
U.S.S.R.

Alan Shuchat (*Translator*)
Department of Mathematics
Wellesley College
Wellesley, MA 02181
U.S.A.

Library of Congress Cataloging-in-Publication Data
Gindikin, S. G. (Semyon Grigor'evich)
   Tales of physicists and mathematicians.
   Translation of: Rasskazy o fizikakh i matematikakh.
   Bibliography: p.
   1. Physics—History.   2. Mathematics—History.
3. Science—History.   I. Title.
QC7.G5613   1988   530'.09   87-24971

   CIP-Kurztitelaufnahme der Deutschen Bibliothek
   Tales of physicists and mathematicians / Semyon
   Grigorevich Gindikin. Transl. by Alan Shuchat. —
   Boston ; Basel : Brikhäuser, 1988.
      Einheitssacht.: Rasskazy o fizikakh i
      matematikakh ⟨engl.⟩

      ISBN-13: 978-1-4612-8409-3   e-ISBN-13: 978-1-4612-3942-0
      DOI: 10.1007/978-1-4612-3942-0

   NE: Gindikin, Semen G. [Hrsg.]; EST

Original Russian edition: *Rasskazy o fizikakh i matematikakh.*
   Quant Library, vol. 14, Moscow: Nauka; first edition:
                1981, second edition: 1985.

© Birkhäuser Boston, 1988
Softcover reprint of the hardcover 1st edition 1988

Typeset by Best-set Typesetter Ltd., Hong Kong.

9 8 7 6 5 4 3 2 1

Contents

From the Foreword to the Russian Edition

This book is based on articles published in *Quant* Magazine over the course of several years. This explains a certain element of randomness in the choice of the people and events to which the stories collected in the book are devoted. However, it seems to us that the book discusses principal events in the history of science that deserve the attention of devotees of mathematics and physics.

We cover a time span of four centuries, beginning with the sixteenth. The sixteenth century was a very important one for European mathematics, when its rebirth began a thousand years after the decline of ancient mathematics. Our story begins at the very moment when, after a 300-year-long apprenticeship, European mathematicians were able to obtain results unknown to the mathematicians of either ancient Greece or the East: they found a formula for the solution of the third-order polynomial equation. The events of the next series of tales begin at the dawn of the seventeenth century when Galileo, investigating free fall, laid the foundation for the development of both the new mechanics and the analysis of infinitely small quantities. The parallel formulation of these two theories was one of the most notable scientific events of the seventeenth century (from Galileo to Newton and Leibniz). We also tell of Galileo's remarkable astronomical discoveries, which interrupted his study of mechanics, and of his dramatic struggle on behalf of the claims of Copernicus. Our next hero, Huygens, was Galileo's immediate scientific successor. The subject we take is his work over the course of forty years to create and perfect the pendulum clock. A significant part of Huygens' achievements in both physics and mathematics was directly stimulated by this activity. The seventeenth century is also represented here by Pascal, one of the most surprising personalities in human history. Pascal began as a geometer, and his youthful work signified that European mathematics was already capable of competing with the great Greek mathematicians on their own territory — geometry. A hundred years had passed since the first successes of European mathematics in algebra.

Towards the end of the eighteenth century, mathematics unexpectedly found itself with no fundamental problems on which the leading scholars would otherwise have concentrated their efforts. Some approximation of mathematical analysis had been constructed; neither algebra nor geometry had brought forth suitable problems up to that time. Celestial mechanics "saved" the day. The greatest efforts of the best mathematicians, beginning with Newton, were needed to construct the theory of motion of heavenly bodies, based on the law of universal gravitation. For a long time, almost all good mathematicians had considered it a matter of honor to demonstrate their prowess on some problem of celestial mechanics. Even Gauss, to whom the last part of this book is devoted, was no exception. But Gauss came to these problems as a mature scholar, and instead made his debut in an unprecedented way. He solved a problem that had been outstanding for 2000 years: He proved it was possible to construct a regular 17-gon with straightedge and compass. The ancient Greeks had known how to construct regular n-gons for $n = 2^k$, $3 \cdot 2^k$, $5 \cdot 2^k$, and $15 \cdot 2^k$, and had spent much energy on unsuccessful attempts to devise a construction for other values of n. From a technical point of view, Gauss' discovery was based on arithmetical considerations. His work summed up a century and a half of converting arithmetic from a collection of surprising facts about specific numbers, accumulated from the deep past, into a science. This process began with the work of Fermat and was continued by Euler, Lagrange, and Legendre. It was startling that the young Gauss, with no access to the mathematical literature, independently reproduced most of the results of his great predecessors.

Observing the history of science from points chosen more or less at random turns out to be instructive in many ways. For example, numerous connections revealing the unity of science in space and time come into view. Connections of a different kind are revealed in the material considered in this book: the immediate succession from Galileo to Huygens, Tartaglia's ideas on the trajectory of a projectile brought by Galileo to a precise result, Galileo's profiting from Cardano's proposal for using the human pulse to measure time, Pascal's problems on cycloids being opportune for Huygens' work on the isochronous pendulum, the theory of motion of Jupiter's moons, which were discovered by Galileo, to which scholars of several generations tried to make some small contribution, etc.

One can note many situations in the history of science that repeat often with small variations (in the words of the French historian de Tocqueville, "history is an art gallery with few originals and many copies"). Consider, for example, how the evaluation of a scientist changes over the centuries. Cardano had no doubt that his primary merit lay in medicine and not in mathematics. Similarly, Kepler considered his main achievement to be the "discovery" of a mythical connection between the planetary orbits and the regular polyhedra. Galileo valued none of his discoveries more than the erroneous assertion that the tides prove the true motion of the earth (to a significant extent, he sacrificed his well-being for the sake of its publication). Huygens considered his most important result to be the application of the cycloid pendulum to clocks, which turned out to be completely useless in practice, and Huygens could have considered himself generally unsuccessful since he could not solve his greatest problem — to construct a naval chronometer (much of what is considered today to be his fundamental contribution was only a means for constructing naval chronometers). The greatest people are defenseless against errors of prognosis. In fact, a scientist sometimes makes the critical decision to interrupt one line of research in favor of another. Thus, Galileo refused to carry through to publication the results of his twenty-year-long work in mechanics, first being diverted for a year to make astronomical observations and then essentially ceasing scientific research, in the true sense of the word, for twenty years in order to popularize the heliocentric system. A century and a half later, Gauss' work on elliptic functions remained unpublished, again for the sake of astronomy. Probably neither foresaw how long the interruption would be, and neither saw around him anyone who could have threatened his priority. Galileo succeeded in publishing his work in mechanics, after 30 years(!), when the verdict of the Inquisition closed off for him the possibility of other endeavors. Only a communication by Cavalieri about the trajectory of a projectile being parabolic forced Galileo to worry a bit, although it did not encroach on his priority. Gauss did not find time to complete his results, also for thirty years, and they were rediscovered by Abel and Jacobi.

The selection of material and the nature of its presentation were dictated by the fact that the book and the articles on which it is based are addressed to lovers of mathematics and physics and, most of all, to students. We have always given priority to a precise account of specific scientific achievements (Galileo's work in mechanics, Huygens' mathematical and mechanical research in connection with pendulum clocks, and Gauss' first two mathematical works). Unfortunately, this is not always possible, even with ancient works. There is no greater satisfaction than following the flight of fancy of a genius, no matter how long ago he lived. It is not only a matter of this being beyond the *amateur* in the case of contemporary works. To be able to feel the revolutionary character of an achievement of the past is an important part of culture.

We wish to stress that the tales collected in this book do not have the nature of texts in the history of science. This is revealed in the extensive adaptation of the historical realities. We freely modernize the reasoning of our scientists: We use algebraic symbols in Cardano's proofs, we introduce free-fall acceleration in Galileo's and Huygens' calculations (in order not to bother the reader with endless ratios), we work with natural logarithms instead of Naperian ones in the story of Napier's discovery, and we use Galileo's latest statements in order to reconstruct the logic of his early studies in mechanics. Throughout, we consciously disregard details that are appropriate for a work in the history of science in order to present vividly a small number of fundamental ideas.

In this edition, corrections of an editorial nature have been made and any inaccuracies and misprints that were noted have been eliminated.

Translator's Note

Wherever possible, citations have been made to English versions of the works discussed in the book. In addition, since many of the quotations that appear were taken from various European languages (including English), I have tried to use existing translations or work directly from the original.

In the Soviet literature, it is common for books such as this to lack the detailed references present in more scholarly works. As a result, it has been difficult to locate the sources of many quotations. Some have thus been translated twice, first into Russian and then into English, and inaccuracies may have crept in. There is a story (apocryphal?) about a computer that translated "the spirit is willing but the flesh is weak" into Russian and back again, ending up with "the wine is strong but the meat is rancid". I trust these results are more palatable!

A.S.

Ars Magna (*The Great Art*)

In 1545 a book appeared, by Gerolamo Cardano, whose title began with these words, *Ars Magna* in Latin. It was essentially devoted to solving third- and fourth-order equations, but its value for the history of mathematics far surpassed the limits of this specific problem. Even in the twentieth century, Felix Klein, evaluating this book, wrote, "This work, which is of the greatest value, contains the germ of modern algebra, surpassing the bounds of ancient mathematics."

The sixteenth century was the century in which European mathematics was revived after the hibernation of the Middle Ages. For a thousand years the work of the great Greek geometers was forgotten, and in part irrevocably lost. From Arab texts, the Europeans learned not only about the mathematics of the East but also about ancient mathematics. It is characteristic that in the spread of mathematics across Europe a major role was played by traders, for whom journeys were a means of both obtaining information and spreading it. The figure of Leonardo of Pisa (1180–1240), better known as Fibonacci (son of Bonacci), especially stands out. His name is immortalized by a remarkable numerical sequence (the Fibonacci numbers). Science can lose its royal status very quickly and centuries may be needed to reestablish it. For three centuries European mathematicians remained apprentices, although Fibonacci undoubtedly did some interesting work. Only in sixteenth century Europe did significant mathematical results appear that neither the ancient nor the Eastern mathematicians knew. We are talking about the solution of third- and fourth-degree equations.

Typically, the achievements of the new European mathematics were in algebra, a new field of mathematics that arose in the East and was essentially taking only its first steps. For at least a hundred years, it would be beyond the power of the European mathematicians not only to achieve something in geometry comparable to the great geometers, Euclid, Archimedes, and Apollonius, but even to master fully their results.

Legend ascribes to Pythagoras the phrase "All is number." But after Pythagoras, geometry gradually came to dominate all of Greek mathematics. Euclid even put the elements of algebra into geometric form. For example, a square was divided by lines parallel to its sides into two smaller squares and two equal rectangles. The formula $(a + b)^2 = a^2 + b^2 + 2ab$ was obtained by comparing areas. But to be sure, there was no algebraic notation at the time, and expressing the result in terms of areas was definitive. Mathematical statements were very awkward. In essence, construction problems with straightedge and compass led to solving quadratic equations and to considering expressions that contained square roots (quadratic irrationals). For example, Euclid considered expressions of the form

$$\sqrt{(a + \sqrt{b})}$$

in detail (in different language). To a certain extent, the Greek geometers understood the link between the classical unsolved construction problems (duplicating a cube and trisecting an angle) and cubic equations.

With the Arab mathematicians, algebra gradually became distinct from geometry. However, as we will see below, the solution of the cubic equation was obtained by geometric means (the debut of algebraic formulas for solving even the quadratic equation came only with Bombelli in 1572). The algebraic assertions of the Arab mathematicians are stated as recipes for the solution of one-of-a-kind arithmetic problems, usually of an "everyday" sort (for example, dividing an inheritance). Rules are formulated for specific examples but so that similar problems can be solved. Until recently rules for solving arithmetic problems (the rule of three,[1] etc.) were sometimes stated this way. Stating rules in general form almost inevitably requires a developed symbolism, which was still far off. The Arab mathematicians did not go further than solving quadratic equations and some specially chosen cubics.

Problems involving cubic equations disturbed both the Arab mathematicians and their European apprentices. A surprising result in this direction belongs to Leonardo of Pisa. He showed that the roots of the equation $x^3 + 2x^2 + 10x = 20$ cannot be expressed through Euclidean irrationals of the form

$$\sqrt{(a + \sqrt{b})}.$$

This statement is startling for the beginning of the thirteenth century and foreshadows the problem of solving equations in radicals, which was thought of significantly later. Mathematicians did not see the path that led to solving the general cubic equation.

The state of mathematics at the turn of the sixteenth century was summed up by Fra Luca Pacioli (1445–1514) in his book, *Summa de Arithmetica* (1494), one of the first printed mathematics books and written in

---

[1] a mechanical way of solving proportion problems—*Transl.*

Italian rather than Latin.[2] At the end of the book he states that "the means [for solving cubic equations] by the art of algebra are not yet given, just as the means for squaring the circle are not given." The comparison sounds impressive, and Pacioli's authority was so great that most mathematicians (even including our heroes at first, as we shall see) believed that the cubic equation could not be solved in general.

Scipione dal Ferro

There was a man who was not deterred by Pacioli's opinion. He was a professor of mathematics in Bologna named Scipione dal Ferro (1465–1526), who found a way to solve the equation

$$x^3 + ax = b. \qquad (1)$$

Negative numbers were not yet in use and, for example,

$$x^3 = ax + b \qquad (2)$$

was thought of as completely different! We have only indirect information about this solution. Dal Ferro told it to his son-in-law and successor on the faculty, Annibale della Nave, and to his student Antonio Maria Fior. The latter decided, after his teacher's death, to use the secret confided to him to become invincible in the problem-solving "duels" that were then quite widespread. On February 12, 1535, Niccolò Tartaglia, one of the major heroes of our story, nearly became his victim.

Niccolò Tartaglia

Tartaglia was born around 1500 in Brescia, into the family of a poor mounted postman named Fontana. During his childhood, when his native city was captured by the French, he was wounded in the larynx and spoke with difficulty thereafter. Because of this he was given the nickname "Tartaglia" (stutterer). Early on he came under the influence of his mother, who tried to enroll him in school. But the money ran out when the class reached the letter "k," and Tartaglia left school without having learned to write his own name. He continued to study on his own and became a "master of the abacus" (something like an arithmetic teacher) in a private business college. He traveled a lot throughout Italy until landing in Venice in 1534. Here his scientific studies were stimulated by contact with engineers and artillerymen of the famed Venetian arsenal. They asked Tartaglia, for example, at what angle to aim a gun so that it shoots the farthest. His answer, a 45° angle, surprised his questioners. They did not believe that they had to raise the barrel so high, but "several private experiments" proved he was right. Although Tartaglia said he had

---

[2] despite its title—*Transl.*

"mathematical reasons" for this assertion, it was more of an empirical observation (Galileo gave the first proof).

Tartaglia published two books, one a sequel of the other: *La Nuova Scientia (The New Science [of Artillery]*, 1537) and *Quesiti et Inventioni Diverse (Problems and Various Inventions*, 1546), where the reader is promised "...new inventions, stolen from neither Plato nor Plotinus, nor from any other Greek or Roman, but obtained only by art, measurement, and reasoning." The books were written in Italian, in the form of a dialogue, which was later adopted by Galileo. In several respects, Tartaglia was Galileo's predecessor. Although in the first of these books he followed Aristotle in saying that a projectile launched at an angle first flies along an inclined straight line, then along a circular arc, and finally falls vertically, in the second book he wrote that the trajectory "does not have a single part that is perfectly straight." Tartaglia was interested in the equilibrium of bodies on an inclined plane and in free fall (his student Benedetti convincingly showed that the behavior of a falling body does not depend on its weight). Tartaglia's translations of Archimedes and Euclid into Italian and his detailed commentaries played an important role (he called Italian a "national" language, as opposed to Latin). In his personal qualities Tartaglia was far from irreproachable and was very difficult in interpersonal relations. Bombelli (who was admittedly not impartial; more on him later) wrote that "this man was by nature so inclined to speak badly, that he took any sort of abuse as a compliment." According to other information (Pedro Nuñes) "he was at times so excited that he seemed mad."

Let us return to the duel before us. Tartaglia was an experienced combatant and hoped to win an easy victory over Fior. He was not frightened even when he discovered that all thirty of Fior's problems contained equation (1), for various values of a and b. Tartaglia thought that Fior himself could not solve these problems, and hoped to unmask him: "I thought that not a single one could be solved, because Fra Luca [Pacioli] assures us of their difficulty, that such an equation cannot be solved by a general formula." After fifty days, Tartaglia was supposed to submit the solution to a notary. When the time limit had almost elapsed, he heard a rumor that Fior had a secret method for solving equation (1). He was not pleased by the prospect of hosting a victory meal for Fior's friends, one friend for each problem the victor solved (those were the rules!). Tartaglia put forth a titanic effort, and fortune smiled on him eight days before the deadline of February 12, 1535: He found the method he had hoped for! He solved all the problems in two hours. His opponent did not solve a single one of the problems Tartaglia had given him. Strangely enough, Fior could not handle one problem that could be solved by dal Ferro's formula (Tartaglia had posed it with a certain trick in mind for solving it), but we will see that the formula is not easy to use. Within a day Tartaglia found a method for solving equation (2).

Niccolò Tartaglia (only known portrait), 1500(?)–1557.

Many people knew about the Tartaglia–Fior duel. In this situation a secret weapon could not help but could rather hurt Tartaglia in further duels. Who would agree to compete with him if the outcome were predetermined? All the same, Tartaglia turned down several requests to reveal his method for solving cubic equations. But one who made the request achieved his goal. This was Gerolamo Cardano, the second hero of our tale.

Gerolamo Cardano

He was born in Pavia on September 24, 1501. His father, Fazio Cardano, an educated lawyer with broad interests, was mentioned by Leonardo da Vinci. Fazio was his son's first teacher. After graduating from the University of Padua, Gerolamo decided to devote himself to medicine. But he was an illegitimate child and so was denied admission to the College of Physicians in Milan. Cardano practiced in the provinces for a long time until August 1539, when the college admitted him anyway, specially changing the rules to do so. Cardano was one of the most famous doctors of his time, probably only second to his friend Andreas Vesalius. In his declining years Cardano wrote his autobiography, *De Vita Propria Liber* (*The Book of My Life*). It contains recollections of his mathematical work, as well as detailed descriptions of his medical research. He claimed that he prescribed cures for up to 5000 difficult diseases and solved some 40,000 problems and questions, as well as up to 200,000 smaller ones. Of course these figures should be taken with a large dose of skepticism, but Cardano was undoubtedly a famous physician. He described cases from his medical practice where he applied pressure in order to cure noted personalities (Archbishop James Hamilton of Scotland, Cardinal Morone, etc.), claiming that he had only three failures. In modern terms he was evidently an outstanding diagnostician, but he did not pay great attention to anatomical information, unlike Leonardo da Vinci and Vesalius. In his autobiography Cardano places himself alongside Hippocrates, Galen, and Avicenna (the latter's ideas were especially close to his own).

However his medical studies did not fill up Cardano's time. In his free moments he studied everything under the sun. For example, he constructed horoscopes for persons living and dead (Christ, King Edward VI of England, Petrarch, Dürer, Vesalius, and Luther). These studies harmed his reputation among his successors (according to one unkind legend, Cardano committed suicide in order to verify his own horoscope). But we must remember that at that time astrology was completely respectable (astronomy was a part of astrology—natural astrology as opposed to the

Gerolamo Cardano (at the age of 68), 1501–1576.

astrology of predictions). The pope himself utilized the work of Cardano the astrologer.

In his scientific activities Cardano was an encyclopedist, but a lone encyclopedist, which was typical for the time of the Renaissance. Only after a century and a half did the first academies appear, in which scholars specialized in more or less narrow fields. Real encyclopedias could only be created with such collaborative efforts. The lone encyclopedist was in no position to verify much of the information he was given. In Cardano's case a large role was played by the peculiarities of his personality and his psychological bent. He believed in magic, premonitions, demons, and in his own supernatural ability. He describes in detail the events that convinced him of this (there was no bleeding in any collision he saw, from neither people nor animals, even in hunting; he learned in advance, from signs, about the events leading up to his son's death, etc.). Cardano believed he possessed a gift of vision (a "harpocratic" feeling) that allowed him to divine both an inflamed organ in an ill patient and the fall of the dice in a game of chance, and to see the mark of death on an interlocutor's face. Dreams, which he remembered in the finest detail and carefully described, played a great role in his life. Contemporary psychiatrists have used these descriptions to try to determine his disease. Cardano writes that constantly recurring dreams, together with the desire to immortalize his name, were his main reasons for writing books. In his encyclopedias *De Subtilitate Rerum (On Subtlety)* and *De Rerum Varietate (On a Variety of Matters)*, he again gives a lot of space to descriptions of the author and his father.

But these books also contain many personal observations and carefully digested communications from others. His readiness to discuss fantastic theories and his peculiar credulity do not play only a negative role. Thanks to them, he discusses things that his more careful colleagues decided to speak of only many years later (see below about complex numbers). It doesn't always pay to follow authority. It is not clear how familiar Cardano was with the works of Leonardo da Vinci (this also applies to other sixteenth century Italian authors; Leonardo became widely known only at the very end of the eighteenth century). *De Subtilitate Rerum* was brought to France and served as a popular textbook on statics and hydrostatics throughout the seventeenth century. Galileo employed Cardano's instructions for using the human pulse to measure time (in particular, for observing the oscillations of the cathedral chandelier). Cardano asserted that perpetual motion is impossible, some of his remarks can be interpreted as the principle of virtual displacements (according to Pierre Duhem, the well-known historian of physics), and he studied the expansion of steam. Cardano adhered to the theory, first conceived of in the third century B.C., that explained the tides by the motion of the moon and sun. He was the first to explain clearly the difference between magnetic and electrical attraction (we have in mind the type of phenomenon observed as early as Thales (c.640–c.546 B.C.), such as the attraction of straw to polished amber).

Cardano was no stranger to experimental research either, nor to the construction of practical devices. In his declining years he established experimentally that the ratio of the density of air to water is 1/50. In 1541, when King Charles V of Spain conquered Milan and entered the city in triumph, Cardano, as Rector of the College of Physicians, walked alongside him near the baldachin (canopy). In response to the honor shown him, he offered to supply the royal team with a suspension from two shafts, which would keep the coach horizontal when it rocked (roads at the time were long and bad). Such a system is now called a Cardan suspension (Cardan shaft, Cardan joint) and is used in automobiles. The truth requires us to note that the idea of such a system arose in antiquity and that, at the very least, there is a drawing of a ship's compass with a Cardan suspension in Leonardo da Vinci's *Codice Atlantico*. Such compasses became common during the first half of the sixteenth century, obviously without Cardano's influence.

Cardano wrote a great many books, of which some were published, some remained in manuscript, and some were destroyed by him in Rome in anticipation of arrest. His voluminous book, *De Libris Propriis (On My Own Books)*, contained only a description of the books he had written. His books on philosophy and ethics were popular for many years, and *On Consolation* was translated into English and influenced Shakespeare. Some Shakespeare-philes even claim that Hamlet speaks his monologue "To be or not to be..." while holding this book in his hands.

Much can be said about Cardano's personality. He was passionate, quick-tempered, and often played games of chance. Cardano gambled at chess for forty years ("I could never express in a few words how much damage this caused my home life, without any compensation") and at dice for twenty-five ("but dice harmed me even more than chess"). From time to time he threw away his studies for gambling, and fell into unpleasant situations. A collateral product of this passion of Cardano's was *Liber de Ludo Aleae (The Book on Games of Chance)*, written in 1526 but published only in 1663. This book contains the beginnings of probability theory, including a preliminary statement of the law of large numbers, some combinatoric questions, and observations on the psychology of gamblers.

Here are a few words about Cardano's nature. He himself writes: "This I recognize as unique and outstanding among my faults—the habit, which I persist in, of preferring to say above all things what I know to be displeasing to the ears of my hearers. I am aware of this, yet I keep it up wilfully.... And I have made many, nay, numberless blunders, wherever I wished to mingle with my fellows....I blundered, almost unavoidably, not solely because of lack of deliberations, and an ignorance of...manners and cusoms, but because I did not duly regard certain of those conventions which I learned about long afterwards, and with which cultivated men, for the most part, are acquainted."[3] For friends and students he could be yet another person. Bombelli wrote that Cardano had "a more god-like than human appearance."

Cardano and Tartaglia

Towards 1539, Cardano was completing his first mathematical book, *Practica Arithmeticae Generalis*, envisioned to replace Pacioli's book. Cardano burned with desire to adorn his book with Tartaglia's secret. At his request, the bookseller Zuan Antonio da Bassano met with Tartaglia in Venice on January 2, 1539. He asked Tartaglia, in the name of "a worthy man, physician of Milan, named Messer Gerolamo Cardano," to give him the rule for solving equation (1), either to publish in the book or under promise to keep it secret. The response was negative: "Tell his Excellency that he must pardon me, that when I publish my invention it will be in my own work and not in that of others..."[4] Tartaglia also refused to communicate the solutions to Fior's thirty problems and only stated the questions (which could have been obtained from the notary), and refused to solve seven problems sent by Cardano. Tartaglia suspected that Cardano was a straw man for the mathematician Zuanne de Tonini da Coi, who had long been trying unsuccessfully to learn the secret.

On February 12th, Cardano sent Tartaglia comments about his book *La Nuova Scientia* and repeated his requests. Tartaglia was implacable, agreeing to solve only two of Cardano's problems. On March 13th Cardano invited Tartaglia to visit him, expressed interest in his artillery instruments, and promised to present him to the Marchese del Vasto, the Spanish governor of Lombardy. Evidently, this perspective enticed Tartaglia, since he accepted the invitation and the critical meeting took place on March 25th at Cardano's home.

Here is an excerpt from the notes of this meeting (one must keep in mind that the record was made by Tartaglia; Ferrari, Cardano's student, claimed that it does not completely correspond to the facts):

"**Niccolò**. I say to you: I refused you not just because of this one chapter and the discoveries made in it, but because of those things that could be discovered knowing it, since this is the key that unlocks the way to the study of countless other areas. I would have long ago found a general rule for many other problems, if I had not at present been occupied with translating Euclid into the national language (I have now brought the

---

[3] From *The Book of My Life* by Jerome Cardan (Gerolamo Cardano), trans. Jean Stoner. Copyright 1930 by E.P. Dutton. Reprinted by permission of the publisher, E.P. Dutton, a division of NAL Penguin Inc.

[4] Oystein Ore, *Cardano, the Gambling Scholar*. Copyright 1953, © renewed 1981 by Princeton University Press. Reprinted with permission of Princeton University Press.

translation up to Book XIII). But when this task, which I have already begun, is done, I plan to publish the work for practical application together with a new algebra....If I give it to some theorist (such as your Excellency), then he could easily find other chapters with the help of this explanation (for it is easy to apply this explanation to other questions) and publish the fruit of my discovery under another name. All my plans would be ruined by this.

Messer Gerolamo. I swear to you by the Sacred Gospel, and on my faith as a gentleman, not only never to publish your discoveries, if you tell them to me, but I also promise and pledge my faith as a true Christian to put them down in cipher so that after my death no one shall be able to understand them. If I, in your opinion, am trustworthy then do it, and if not then let us end this conversation.

Niccolò. If I did not believe an oath such as yours then, of course, I myself would deserve to be considered an atheist."

Thus, Tartaglia convinced himself. He communicated his solution in the form of a Latin poem. Isn't it true that it is hard to understand from these notes *what* induced Tartaglia to change his decision? Was he really shaken by Cardano's vow? What happened later is not well understood. Having revealed his secret, the uneasy Tartaglia left immediately, refusing to meet the marchese for whom he had undertaken the journey. Could Cardano have hypnotized him? In all likelihood, Tartaglia's account is inaccurate.

Tartaglia was somewhat reassured when on May 12th he received the *Practica Arithmeticae Generalis*, freshly printed, without his recipe. In an accompanying letter, Cardano wrote: "I have verified the formula and believe it has broad significance."

Cardano received from Tartaglia a ready-to-use method for solving equation (1), without any hint of proof. He spent a great deal of effort on carefully verifying and substantiating the rule. From our standpoint it is not easy to understand the difficulty: Just substitute into the equation and verify it! But the absence of a well-developed algebraic notation made what any schoolchild today can do automatically accessible only to a select few. Without knowing the original texts from that time we cannot appreciate how much the algebraic apparatus "economizes" thought. The reader must always keep this in mind, so as not to be deluded by the "triviality" of the problems over which passions seethed in the sixteenth century.

Cardano spent years of tense work, trying to understand thoroughly the solution of cubic equations. He obtained a recipe (after all, they didn't know how to write formulas!) for solving equations (1) and (2), as well as

$$x^3 + b = ax \tag{3}$$

and equations containing x^2. He certainly "outstripped" Tartaglia. All this happened against the background of a consolidation of Cardano's position; in 1543 he became professor at Pavia. "My knowledge of astrology," wrote Cardano, "led me to the conclusion that I would not live more than forty years and, in any case, would not reach the age of forty-five....The year arrived that was supposed to be the last one of my life and that, on the contrary, turned out to be its beginning—namely, the forty-fourth."

Luigi Ferrari

For some time, Cardano had been assisted in his mathematical work by Luigi Ferrari (1522–1565). In a list Cardano made of his fourteen students, Ferrari appears as the second chronologically and one of the three most outstanding. Cardano, believing in signs, wrote that on November 14, 1536, when the fourteen-year-old Luigi and his brother arrived in Bologna, "a magpie in the courtyard chirred for such an unusually long time that we all expected someone to arrive." Ferrari was a man of phenomenal ability. He had such a stormy temper that even Cardano was sometimes afraid to speak with him. We know that at seventeen, Ferrari returned from a brawl without a single finger on his right hand. He was unreservedly devoted to his teacher and for a long time was his secretary and confidant. Ferrari's contribution to Cardano's mathematical work was quite substantial.

In 1543 Cardano traveled with Ferrari to Bologna, where della Nave allowed him to examine the papers of the late dal Ferro. They became convinced that dal Ferro had known Tartaglia's rule. It is interesting that they evidently knew almost nothing about dal Ferro's formula. Cardano would hardly have pursued Tartaglia so energetically had he known that the same information could have been obtained from della Nave (he had not consulted him before 1543). Almost everyone now agrees that dal Ferro had the formula, that Fior knew it, and that Tartaglia rediscovered it knowing that Fior had it. However not one of the steps in this chain has been strictly proven! Cardano spoke of it, but Tartaglia wrote at the end of his life: "...I can testify that the theorem described was not proved before by Euclid or by anyone else but only by one Gerolamo Cardano, to whom we showed it....In 1534 [elsewhere February 4, 1535—S.G.] in Venice, I found a general formula for the equation..." It is hard to untangle this confused story.

Ars Magna

Either familiarity with dal Ferro's papers, strong pressure from Ferrari, or, most likely, an unwillingness to bury the results of many years' work led Cardano to include everything he knew about cubic equations in his book, *Artis Magnae Sive de Regulis Algebraicis* (*The Great Art, or The Rules of Algebra*), which appeared in 1545. It has come to be called simply *Ars Magna* (*The Great Art*).

At the beginning, Cardano lays out the history of the problem: "... In our own days Scipione del Ferro of Bologna has solved the case of the cube and first power equal to a constant, a very elegant and admirable accomplishment. Since this art surpasses all human subtlety and the perspicuity of mortal talent and is a truly celestial gift and a very clear test of the capacity of men's minds, whoever applies himself to it will believe that there is nothing that he cannot understand. In emulation of him, my friend Niccolò Tartaglia of Brescia, wanting not to be outdone, solved the same case when he got into a contest with his [Scipione's] pupil, Antonio Maria Fior, and, moved by my many entreaties, gave it to me. For I had been deceived by the words of Luca Paccioli, who denied that any more general rule could be discovered than his own. Notwithstanding the many things which I had already discovered, as is well known, I had despaired and had not attempted to look any further. Then, however, having received Tartaglia's solution and seeking for the proof of it, I came to understand that there were a great many other things that could also be had. Pursuing this thought and with increased confidence, I discovered these others, partly by myself and partly through Lodovico Ferrari, formerly my pupil."[5]

In modern form, the method by which Cardano solved equation (1) can be presented in the following way. We will seek a solution to (1) in the form $x = \beta - \alpha$. Then $x + \alpha = \beta$ and

$$x^3 + 3x^2\alpha + 3x\alpha^2 + \alpha^3 = \beta^3. \tag{4}$$

Since $3x^2\alpha + 3x\alpha^2 = 3x\alpha(x + \alpha) = 3x\alpha\beta$, we can rewrite (4) as

$$x^3 + 3\alpha\beta x = \beta^3 - \alpha^3. \tag{5}$$

Let us try to choose the pair (α, β) in terms of (a, b) so that (5) coincides with (1). In order for this to happen, (α, β) must be a solution of the system

$$3\alpha\beta = a, \qquad \beta^3 - \alpha^3 = b$$

or equivalently,

$$\beta^3(-\alpha^3) = -\frac{a^3}{27} \qquad \beta^3 + (-\alpha^3) = b.$$

By Vieta's theorem[6], β^3 and $-\alpha^3$ will be roots of the auxiliary quadratic equation $y^2 - by - a^3/27 = 0$. Since we are seeking *positive* roots of (1), $\beta > \alpha$. This means that

$$\beta^3 = \frac{b}{2} + \sqrt{\frac{b^2}{4} + \frac{a^3}{27}}, \qquad -\alpha^3 = \frac{b}{2} - \sqrt{\frac{b^2}{4} + \frac{a^3}{27}}.$$

---

[5] Girolamo Cardano, *The Great Art, or The Rules of Algebra*, trans. T. Richard Witmer. © 1968 by The M.I.T. Press, pp. 8–9. Here and in other places there are slight variations in names.—*Transl.*
[6] François Viète (or Vieta, 1540–1603) himself lived after Cardano, but Cardano essentially knew this special case, now known as Vieta's theorem, of a result Vieta later proved.

Thus,

$$x = \sqrt[3]{\frac{b}{2} + \sqrt{\frac{b^2}{4} + \frac{a^3}{27}}} - \sqrt[3]{-\frac{b}{2} + \sqrt{\frac{b^2}{4} + \frac{a^3}{27}}}.$$

When a and b are positive, the root x is thus also positive.

The calculation we present here follows only the idea of Cardano's argument. He himself argued geometrically: If we divide a cube of side $\beta = \alpha + x$ by planes, parallel to its faces, into one cube of side α and one of side x, then in addition to these two cubes we obtain three rectangular parallelepipeds with sides α, α, x and three with sides α, x, x. Their volumes are related according to (4), and if we combine the parallelepipeds of different types pairwise, then we obtain (5). "When, moreover, I understood that the rule that Niccolò Tartaglia handed to me had been discovered by him through a geometrical demonstration, I thought that this would be the royal road to pursue in all cases."[7] Cardano may have known of an analogous argument for the quadratic equation, due to al-Khowârizmî.[8]

Equation (2) can be solved using the substitution $x = \beta + \alpha$, but it can happen that the original equation has three real roots while the auxiliary quadratic equation has none. This is called the *irreducible* case, and it gave Cardano (and, probably, Tartaglia) much trouble.

Cardano solved equation (3) by an argument that was daring at the time because it played on the negativity of the root. No one before had used negative numbers so decisively, and even Cardano himself far from used them freely. He considered equations (1) and (2) separately!

Cardano also thoroughly investigated the general cubic equation $x^3 + ax^2 + bx + c = 0$, noting that, in contemporary terms, the substitution $x = y - a/3$ eliminates the x^2 term.

Cardano decided to consider not only negative numbers (he calls them "purely false") but also complex numbers (these he calls "truly sophisticated"). He remarked that if we operate on them according to certain natural rules, then we can ascribe complex roots to a quadratic equation having no real roots. Cardano may have arrived at complex numbers in connection with the irreducible case. (N. Bourbaki, for example, suggests this). If, in this case, we are "undaunted" in carrying out all the operations on the complex numbers that arise during the calculation, then at the end we obtain the correct values of the real roots. But there is no indication whatever that Cardano considered more than quadratic equations here. However, the argument presented for the cubic equation soon appeared— in the hands of Raffael Bombelli (1526–1573), Cardano's successor, a

---

[7] Cardano, *The Great Art*, p. 52.
[8] Mohammed ibn Musa al-Khowârizmî (c.780–c.850), Persian mathematician and astronomer whose name is preserved in the word "algorithm."—*Transl.*

hydraulics engineer from Bologna and the author of the famous *Algebra* (1572).

Cardano understood that the cubic equation $x^3 + ax^2 + bx + c = 0$ can have three real roots and that their sum then equals $-a$. Cardano was unprecedented in making such general assertions. In algebra, as opposed to geometry, practically no proofs were given (traces of this remain today in school mathematics!). Here is yet another of Cardano's observations: If all the terms on the left side of an equation (with positive coefficients) have greater degree than all the terms on the right side, then there is a unique positive root. A whole series of important algebraic concepts comes from *Ars Magna*, e.g., the multiplicity of a root. In general, Cardano's significance in the history of mathematics is determined most of all not by specific achievements (he did not have many) but by the fact that in *Ars Magna* he saw the path along which algebra would develop.

Remarks on Cardano's Formula

Let us analyze the formula as it applies to solving $x^3 + px + q = 0$ over the real numbers. Unlike Cardano, we can allow ourselves to ignore the signs of p and q. Thus,

$$x = \sqrt[3]{-\frac{q}{2} + \sqrt{\frac{q^2}{4} + \frac{p^3}{27}}} + \sqrt[3]{-\frac{q}{2} - \sqrt{\frac{q^2}{4} + \frac{p^3}{27}}}.$$

In calculating x, we must first find the square roots and then the cube roots. We obtain real square roots if $\Delta = 27q^2 + 4p^3 > 0$. The two square root terms, differing by a sign, appear in different summands. Real cube roots are unique, so when $\Delta > 0$ we obtain a unique real value for x.

By studying its graph, it is not hard to see that the cubic trinomial $x^3 + px + q$ in fact has a unique real root when $\Delta > 0$. For $\Delta < 0$ there are three real roots. For $\Delta = 0$ we have one double real root and one single real root, and when $p = q = 0$ we have the triple real root $x = 0$.

We continue with the case $\Delta > 0$ (one real root). It turns out that even if an equation with integer coefficients has an integer root, then calculating the root by the formula can lead to intervening irrational numbers. For example, $x^3 + 3x - 4 = 0$ has the unique real root $x = 1$. For this unique real root, Cardano's formula gives the expression

$$x = \sqrt[3]{2 + \sqrt{5}} + \sqrt[3]{2 - \sqrt{5}}.$$

This means that

$$\sqrt[3]{2 + \sqrt{5}} + \sqrt[3]{2 - \sqrt{5}} = 1.$$

But try to prove this directly! Practically any proof will use the fact that this expression is a root of $x^3 + 3x - 4 = 0$. Perhaps you will find some trick, but straightforward transformations lead to cubic radicals that cannot be removed.

It may be that this explains why Fior could not solve Tartaglia's cubic equation. It probably could have been solved by guessing the answer (this is what Tartaglia had in mind), while dal Ferro's recipe led to intervening irrationals.

The situation is even more confusing in the case of three real roots. This is called the irreducible case. Here $\Delta = 27q^2 + 4p^3 < 0$ and the numbers under the cube root signs are complex. If we find the complex cube roots, then after addition the imaginary parts vanish and we obtain real numbers. But how can we reduce everything to operations on real numbers? For example, finding the square root $\sqrt{a + ib}$ can be reduced to purely real operations on a and b. If that were the case with $\sqrt[3]{a + ib} = u + iv$ then all would be in order. But when we express u and v in terms of a and b, we obtain new cubic equations that again give rise to the irreducible case. We have a vicious circle! In the end, in the irreducible case we cannot express the roots in terms of the coefficients without going beyond the real number system. In this sense the cubic equation with three real roots is unsolvable in radicals over the reals (as opposed to the quadratic equation). This situation does not often receive the attention it deserves.

The Fourth-Degree Equation

Ferrari's personal contribution, the solution of the fourth-degree equation, was also reflected in *Ars Magna.*

In modern terms, Ferrari's method for solving

$$x^4 + ax^2 + bx + c = 0 \tag{6}$$

is as follows (it is easy to reduce the full fourth-degree equation to (6)).

Introducing an auxiliary parameter t, we rewrite (6) in the equivalent form:

$$\left(x^2 + \frac{a}{2} + t\right)^2 = 2tx^2 - bx + \left(t^2 + at - c + \frac{a^2}{4}\right). \tag{7}$$

We now choose a value for t so that the quadratic trinomial (in x) on the right side of (7) has two equal roots. In order for this to happen, its discriminant must be zero:

$$b^2 - 4 \cdot 2t \cdot \left(t^2 + at - c + \frac{a^2}{4}\right) = 0.$$

We have obtained an auxiliary cubic equation in t. Find its root t_0 by Cardano's formula. We can now rewrite (7):

$$\left(x^2 + \frac{a}{2} + t_0\right)^2 = 2t_0\left(x - \frac{b}{4t_0}\right)^2. \tag{8}$$

Equation (8) can be decomposed into a pair of quadratic equations giving the four desired roots.

Thus, by Ferrari's method the fourth-degree equation reduces to solving an auxiliary cubic and two quadratic equations.

Ferrari and Tartaglia

After meeting in 1539, Cardano and Tartaglia rarely corresponded. Once a student told Tartaglia he had heard that Cardano was writing a new book. Tartaglia immediately wrote Cardano a cautioning letter but received a calming answer. Another time Cardano wanted a clarification dealing with the irreducible case, but received nothing substantive in response. It is not hard to imagine the effect on Tartaglia when *Ars Magna* appeared in 1545. In the last part of *Quesiti et Inventioni Diverse* (1546), Tartaglia published the correspondence and notes of discussions dealing with his relations with Cardano, and heaped abuse and rebuke on him. Cardano did not react to this attack, but on February 10, 1547, Ferrari answered Tartaglia. He took exception to Tartaglia's rebukes, pointed out defects in his book, rebuked him in one place for appropriating someone else's result, and found a repetition betraying a bad memory (apparently a serious accusation for the time). Finally, he challenged Tartaglia to a public debate on "Geometry, Arithmetic and the disciplines which depend on them, such as Astrology, Music, Cosmography, Perspective, Architecture, and others."[9] Ferrari was ready to discuss not only what was written in these areas by the Greek, Latin, and Italian authors but even the works of Tartaglia himself, if the latter in turn agreed to discuss the works of Ferrari.

By tradition, such a "cartel" (challenge) required "questions" in response. They appeared on February 19th. Tartaglia wanted to draw Cardano himself into the skirmish: "...I have expressed this in such calumnious and sharp words to incite his Excellency, and not you, to write me in his own hand. I have many accounts to settle with him...." The discussion of the conditions for the duel dragged on. Tartaglia began to understand that Cardano was remaining on the sidelines. Then he started to emphasize Ferrari's lack of independence, calling him Cardano's "creation" or "creature," since Ferrari had called himself that in the first cartel. All questions were addressed to both: "You, Messer Gerolamo, and you, Messer Lodovico...." Much of the correspondence is interesting. For example, the second cartel reproduces a conversation between Cardano and Tartaglia, supposedly overheard by Ferrari: "What more do you want? 'I don't want it divulged,' you say. And why? 'So that no one else shall enjoy my invention.' ...Really, since we are born not for ourselves only but for the benefit of our native land and the whole human race, and when you possess within yourself something good, why don't you want to let others share it?"[10]

The correspondence continued for a year and a half, and suddenly Tartaglia resolutely agreed to a duel in Milan. Why? At the same time, in

---

[9] Ore, p. 88.
[10] Ibid., p. 94.

March 1548, he received a flattering invitation to his native Brescia, to give public lectures (which he had not had occasion to do previously) and to conduct private lessons "in which only certain doctors and people of definite authority will participate." Things had not been going well for him, and there is the opinion that his patrons forced Tartaglia to accept the challenge in hopes that a victory would strengthen his situation. The dispute took place on August 10, 1548, in Milan in the presence of many well-known personalities, including the governor of Milan, but in Cardano's absence. Only Tartaglia's brief notes have been preserved, from which it is almost impossible to recreate the true picture. It seems that Tartaglia sustained a shattering defeat. But don't be mistaken—the debate had no relation to the problem over which the argument had arisen, just as debates as well as physical duels generally have little relation to clarifying the truth. It was hard for the tongue-tied Tartaglia to stand up in public to the sparkling young Ferrari.

The Fate of Our Heroes

Tartaglia was not retained in Brescia, and within a year and a half he returned to Venice without having received even an honorarium for his lectures. His defeat in the debate had hurt him very much. At the end of Tartaglia's life (he died in 1557), *Trattato Generale di Numeri et Misure* (*A General Treatise on Number and Measurement*) began to appear, and its publication was completed only after his death. Very little is said about cubic equations, and no trace of a great treatise on the new algebra, which Tartaglia had spoken about all his life, was discovered in his carefully preserved legacy.

By contrast, Ferrari became very famous after the duel. He gave public lectures in Rome, headed the taxation department in Milan, received an invitation to serve Cardinal Mantui, and took a hand in bringing up the emperor's son. But he left no further trace in science! Ferrari died in 1565 at the age of forty-three; according to legend, he was poisoned by his sister. Speaking of his death, Cardano recalls the lines of Martial, the Roman epigrammatist:

For those without measure life is short, rarely do they attain old age. Whatever you may love or desire, let not your indulgence go beyond bounds.[11]

Cardano outlived them both, but the end of his life was not easy. One of his sons (the doctor Giambattista, on whom Cardano had placed his greatest hopes) poisoned his wife out of jealousy and was executed in 1560. Cardano did not recover from this blow for a long time. Another son, Aldo, became a criminal and robbed his own father. In 1570, Cardano himself was sentenced to prison and his property was confiscated. The

---

[11] Cardano, *My Life*, p. 144; Ore, p. 82.

reason for his arrest is unknown, but the initiative may have come from the Inquisition. While waiting for arrest, Cardano destroyed 120 of his own books. He ended his days in Rome, a "private person" (his expression) receiving a modest pension from the pope. Cardano devoted the last year of his life to his autobiography, *De Vita Propria Liber* (*The Book of My Life*). The last item it mentions is dated April 28, 1576, and on September 21st Cardano died.

In his autobiography, Cardano recalls Tartaglia four times. In one place he approvingly cites his thought that "no one knows everything, and moreover knows nothing that he suspects many people do not know." Elsewhere he says that Tartaglia preferred him as "a rival and victor, and not a friend and a man obliged to do him well." Still, Tartaglia turns out to be among Cardano's critics who "did not go beyond grammar." Finally, on the very last pages, we read: "I confess that in mathematics I received a few suggestions, but very few, from brother Niccolò." It seems that Cardano's soul was uneasy!

Epilogue

The Cardano–Tartaglia problem lay forgotten for a long time. The formula for solving the cubic equation was associated with *Ars Magna* and gradually became known as *Cardano's formula*, although at some time dal Ferro's name was involved (his authorship was stressed by Cardano himself). Such errors in appropriating names are not rare (e.g., recall the axiom of Archimedes, who did not claim this discovery).

The origin of the formula for the cubic equation arose again at the beginning of the nineteenth century. The insulted Tartaglia, who had been practically forgotten, was rediscovered. The almost forgotten story received publicity, and not only professionals, but even amateurs were ready to fight for Tartaglia's honor. The detective aspect of history was very attractive. For how many years should Cardano's promise have been in effect? Was six years really long enough? Why did Tartaglia not publish his formula for ten years? However, as the story spread through the popular literature it became distorted, and Cardano in time turned into an adventurer and villain who stole Tartaglia's discovery and gave it his own name. As we have seen, the situation was more complicated, and such an interpretation, at the very least, oversimplifies the picture.

It was not only a matter of wanting to restore the true picture of the events in a situation where the participants undoubtedly did not tell the whole truth. For many it was important to establish the degree of Cardano's guilt. This question touches on the perennially topical question of proprietary rights in scientific discovery. Concerning today's practice, what strikes us is the difference between the rights of the scientist and of the inventor. The scientist cannot control the future use of his published results but only lay claim to the remembrance of his name. Controlling future use is one of the reasons for keeping inventions secret. At the juncture of the Middle Ages and the Renaissance, mathematical results were kept secret in order to use them in duels.

Towards the end of the nineteenth century, part of the discussion began to take on the character of serious historical-mathematical research. Some original material was published for the first time ("cartels" and "questions"). Mathematicians understood how great a role Cardano's work had played in sixteenth century science. What Leibniz had remarked earlier became clear: "Cardano was a great man for all his faults; without them he would have been perfect."

Moritz Cantor, the great mathematical historian (1829–1920, not to be confused with Georg Cantor, the creator of set theory) and author of a multivolume history of mathematics, held Cardano in very great esteem but not without regret, stating that his human qualities left something to be desired ("genius, but no character"). Cantor proposed, as had Ferrari, that Tartaglia did not rediscover dal Ferro's rule but learned it ready-made and secondhand. He remarked that Tartaglia did not have any significant mathematical works to his credit and, aside from the rule itself and facts that could have been borrowed from *Ars Magna* (which had appeared earlier), his publications and the manuscripts he left contain only elementary remarks about cubic equations. Of course this is no proof, and moreover Tartaglia had many virtues beyond mathematics. Cantor was also suspicious of the fact that Tartaglia's and dal Ferro's solutions were as similar as two drops of water. Gustav Eneström took exception to Cantor and even conducted some sort of research experiment that showed such a coincidence is possible. Ettore Bortoletti did much to explain unclear points, and presented arguments that could confirm a series of seemingly irresponsible statements by Tartaglia.

For a century and a half, passions have calmed down and then heated up again. The desire to obtain a single answer to the question has not died away, but such an answer may simply not exist. As for the formula for the solution to the cubic equation, the name *Cardano's formula* has firmly taken hold.

Appendix: From *De Vita Propria Liber*
(*The Book of My Life*)

Four months before his death Cardano completed his autobiography, which he had anxiously written during the entire previous year and which was supposed to sum up his complex life. He felt death approaching. According to some reports, his personal horoscope associated his end with his seventy-fifth birthday. He died on September 21, 1576[12], a few days

---

[12] Some sources say September 20th, e.g., Ore, p. 23, and Cardano, *My Life*, p. xiii.—*Transl.*

before his birthday. There is a version that he commited suicide in anticipation of his inevitable death or even to confirm the horoscope. In any case, Cardano the astrologist took his horoscope seriously. In his book he described waiting for death at age forty-four, as his earlier horoscope had foretold.

Cardano worried about whether his life had been successful. On the one hand, he lived on a meager papal pension in Rome, in enforced exile from the cities where he had spent the best part of his life, he had recently been in prison, and he was unhappy with his children. On the other hand, Cardano was sure of his own significance. He criticized much from his past, although it is not hard to discover the places where he succeeded in convincing himself that he was right. Cardano's leading idea is the predestination of his life. This is the source of his detailed analysis of the influence of the stars, his association with a "guardian angel," the scrupulous account of signs and omens, and the little events that allow him to build a logically constructed picture of life. In a certain sense, Cardano's aim was, using the scholar's and astrologer's art, to analyze himself in detail as an object of the action of higher powers. A new style was established in science, where conclusions are drawn from the facts as they appear. Therefore Cardano supplies the reader with detailed information about his physical features, drinking patterns, habits, etc., in order for the author and reader to have the same opportunities to draw conclusions. Cardano's book is a remarkable literary monument of the sixteenth century, and allows us to understand much about how one of the wisest men of his time perceived life.

Cardano's book was translated into Russian in 1938 and published by the State Literary Publishing House. Let us see how Cardano talks about himself.[13]

He begins, "This Book of My Life I am undertaking to write after the example of Antoninus the Philosopher, acclaimed the wisest and best of men, knowing well that no accomplishment of mortal man is perfect, much less safe from calumny; yet aware that none, of all ends which man may attain, seems more pleasing, none more worthy than recognition of the truth. No word, I am ready to affirm, has been added to give savor of vainglory, or for the sake of mere embellishment." Cardano describes in detail his native Milan and his ancestors. He says of his birth, "...I was...born on the 24th day of September in the year 1500 [this is evidently a slip of the pen, since Cardano was born in 1501—*S.G.*], when the first hour of the night was more than half run, but less than two-thirds....I was almost dead. My hair was black and curly. I was revived in a bath of warm wine which might have been fatal to any other child." He describes in

---

[13] The following quotations are generally taken from Stoner's English translation of *My Life*, cited earlier.—*Transl.*

detail the positions of Mars, Mercury, and the moon, which presaged that "...consequently I ought to have been a monster, and indeed was so near it that I came forth literally torn from my mother's womb." The "sinister planets" Venus and Mercury foretold that he would be "gifted...with a certain cunning only, and a mind by no means at liberty; my every judgment is, in truth, either too harsh or too forbidding."

In another chapter, he describes his parents: "My father went dressed in a purple cloak, a garment which was unusual in our community; he was never without a small black skullcap. When he talked he was wont to stammer. He was a man devoted to various pursuits. His complexion was ruddy, and he had whitish eyes.... From his fifty-fifth year on he lacked all his teeth. He was well acquainted with the works of Euclid; indeed, his shoulders were rounded from much study." What surprising details! "My mother was easily provoked; she was quick of memory and wit, and a fat, devout little woman."

Later, Cardano gives a short description of his life, after which it is the turn of his discoveries. Here are some details: "I am a man of medium height; my feet are short, wide near the toes, and rather too high at the heels, so that I can scarcely find well-fitting shoes.... My chest is somewhat narrow and my arms slender. The thickly fashioned right hand.... A neck a little long and inclined to be thin, cleft chin, full pendulous lower lip, and eyes that are very small and apparently half-closed; unless I am gazing at something.... My hair and beard were blond.... Old age has wrought changes in this beard of mine, but not much in my hair...," etc. Cardano describes the illnesses he has suffered, and says, "Now I have fourteen good teeth and one which is rather weak; but it will last a long time, I think, for it still does its share." He had ten ailments in all; the tenth was insomnia, which he cured by abstaining from certain kinds of foods.

Cardano tells us that he is timid by nature but gained courage through physical exercise, that he stays in bed for ten hours but sleeps for eight, that he prefers fish to meat and counts off twenty-one kinds of fish that he eats, and that with big fish he eats "the head and belly, but the spine and tail of small ones."

"Vowing to perpetuate my name, I made a plan for this purpose...in some hope of the future, I have scorned the present," we read. Chance, the intrigues of his opponents, and his own astrological finding that he would not live past forty-five interfered with Cardano's aspirations to perpetuate his name. Everything changed when it turned out that the prediction had not come true. Cardano decisively changed his way of life. He delivered lectures early in the morning. "That over, I went walking in the shade beyond the city walls, and later lunched, and enjoyed some music. In the afternoon, I went fishing.... While there I also studied and wrote, and in the evening returned home." Cardano explains why he preferred a career in medicine to law, as his father had wished: "I deemed medicine a profession of sincerer character than law, and a pursuit relying rather upon reason and nature's everlasting law, than upon the opinions of men." He talked about his teaching and debating: "While at the University of Bologna, I usually lectured extemporaneously....Excellent as I may have appeared in these respects, I possessed neither grace in my manner of speech nor talent for making a clever conclusion."

He characteristically lists his virtues: "Notwithstanding, whatsoever my good fortune, or however many the happy issues that attended me, I have never modified my carriage...[nor adopted] a more luxurious mode of dress....To the duties of life I am exceptionally faithful, and particularly in the writing of my books....I have never broken a friendship; neither, if the relationship happened to be discontinued, have I divulged the secrets of my erstwhile friends...." He describes his friends and patrons in detail, but ostentatiously does not enumerate his enemies and rivals. But they repeatedly appear on the pages of the book, even as soon as the next chapter, entitled "Calumny, Defamations, and Treachery of My Unjust Accusers."

Cardano begins with intrigues and experiences some difficulty in choosing examples: He wanted to talk about great and secret intrigues, but intrigues that have already been uncovered cannot be considered secret, and great intrigues are hard to conceal. Philosophizing, he choose the case of obtaining a professorship in Bologna, when it was rumored that he "lectures to the empty benches...that he is a man of bad manners, disagreeable to all, and, in the main, a fool. He is given to disgraceful practices, and he does not show even tolerable skill in medicine...and accordingly has no practice." All this would have been believed if the papal legate in Bologna had not remembered that Cardano had cured his mother. This undermined confidence in the remaining information. Nevertheless the intrigues continued even in Bologna, and Cardano in the end was denied the post, although he reassured himself: "...all this wound up pleasing those who tried so hard for it, but it is not at all to their benefit." Concerning "calumny and lying defamation," Cardano does not dwell on concrete cases, stating that his "calumniators [were] tormented by their own guilty conscience...they have left me more time for collecting my literary works...they have provided me with an opportunity to...devote myself to the investigation of many things not fully revealed to man," and that "I do not hate them."

His pleasures are briefly listed: pen knives (he spent more than twenty gold crowns on them), different types of pens (more than two hundred crowns), precious stones, china, globes of painted glass, rare books, navigation, fishing, the philosophy of Aristotle and Plotinus, the occult, Petrarch's poetry, etc. He preferred solitude to company, not only out of devotion to science but so as not to lose time. We have already mentioned his partiality to gambling with chess and dice.

A separate chapter is devoted to clothing. Cardano finds a description in

Horace that resembles him very much. A rather long discussion with references ends in saying that one must have "four changes of clothes: one warm, one very warm, one light, and one very light. Thus, one gets fourteen different combinations...." He describes his gait, stating that the reason for its unevenness is his constant reflection. He discusses his attitude towards religion and philosophy, stressing the influence of Plato, Aristotle, Plotinus, and especially Avicenna. He lists the "general rules" that he mastered during his life: to thank God and ask Him for help, not to limit oneself to redeeming a loss but always to obtain something in addition, to make the most of time, to respect one's elders, "always to set certainties before uncertainties," "never...to persist willingly in any course which is turning out for the worse," etc. Cardano lists the houses in which he lived, and colorfully describes his poverty and the loss of his father's legacy.

Cardano writes in detail about his wife and children. He writes that he saw his future wife in a dream before meeting her, and that the dream presaged an unhappy marriage. We have already talked about the fate of his children. He describes his travels, usually in connection with his medical activities, and explains the value of the trips.

The longest chapter is devoted to perils and accidents. Cardano describes them in detail, visibly trying to impress the reader to prepare the way for more profound phenomena: "...it is not the significant event which ought to be wondered at, but rather the frequent recurrence of similar instances." Three times, he miraculously escaped danger at nearly the same place: from a tile falling from a wall, from a giant piece of masonry, and from an overturning carriage. Twice he nearly drowned under very romantic circumstances. Cardano was subjected to the attack of a rabid dog, fell into a pit, fell from a carriage onto the road, and was exposed to plague. These accounts read like detective stories. After this comes a sequence of terrible intrigues devised by his competitors, the doctors in Pavia: Here are a scandal involving his daughter's husband, a beam that could have fallen while he entered the Academy, and a poisoning attempt that was averted by the boys who tasted his food. However, everything was unexpectedly ended by the illness or even death of the doctors. In Rome, dangers pursued Cardano because he did not know the streets and "the behavior here [is] so uncouth." But he finally decides that Providence is protecting him, and he stops fearing danger: "Who does not now perceive that all these things have been, as it were, precursors of bliss about to be overtaken...?"

Cardano includes in the book a study of happiness, with examples from his life. He lists the honors shown him, mainly flattering invitations. On the other hand, he recounts unpleasant episodes from his medical practice and discusses their association with dreams. There is an unexpected discussion of the Cardano family coat of arms, to which he decided to add a swallow on the day of his arrest: "I chose the swallow as in harmony...

with my own nature: It is harmless to mankind, it does not shun association with the lowly, and is ever in contact with humankind without becoming familiar. . . ." A questionable comparison! Cardano also lists his teachers and students.

Cardano again discusses his characteristics and surprising events in his life: While a child he had visions of ringlets in the sky, he could not warm up his legs below the knees, and blood did not flow in his presence (he even began to intervene purposely in fights and was never injured); the events leading up to the death of his oldest son; and finally the many dreams which preceded events that later occurred. The descriptions of his dreams are very colorful and detailed.

Later, Cardano lists the ten sciences he was acquainted with in his life and describes forty cases from his medical practice. Then comes a chapter, "Concerning Natural Though Rare Circumstances of My Own Life." Here is the first of these occurrences: ". . . I was born in this century in which the whole world became known, whereas the ancients were familiar with but little more than a third part of it." Also, his house collapsed but his bedroom was spared, his bed caught fire twice, etc. He analyzes in detail the gift for prognostication that constantly appeared in his life, from medicine to card games.

The concluding part of the book again deals with supernatural events, discusses scientific achievements, and lists his books. Cardano again talks about himself, about his guardian angel, lists testimonials to himself, discusses "worldly things," and devotes several pages to maxims for life. Here are some examples: "Friends are your support in adversity, flatterers bring you advice. . . . An illustrious man ought to live under the aegis of a prince. . . . When you wish to wash yourself first prepare a linen towel for wiping. . . . Evil is but a lack of good, and good is of itself a virtue which is within our power to possess, or rather which is indispensable." After these sayings comes "A Lament on the Death of My Son." At the end, we again hear of Cardano's inadequacies, of the changes that come with age, and of the "Quality of Conversation."

Two Tales of Galileo

I. The Discovery of the Laws of Motion

...it was Galileo who laid the first foundations of [dynamics]. Before him, only forces acting on bodies in equilibrium were considered; and although the acceleration of falling bodies and the curved motion of projectiles could only be attributed to the constant force of gravity, no one had been able to determine the laws by which these daily phenomena follow from such a simple cause. Galileo was the first to take this important step, and in so doing began a new and vast arena for the development of mechanics...today, it is the most significant and secure part of this great man's glory. The discoveries of the moons of Jupiter, the phases of Venus, sunspots, etc., only required a telescope and perseverance; however, it needed an extraordinary genius to unravel the natural laws of these phenomena that had always been before everyone's eyes but had nonetheless always escaped the philosophers' reach.

Lagrange[1]

Prologue

Vincenzio Galilei, a well-known Florentine musician, had reflected for a long time on what field to choose for his oldest son Galileo. The son was undoubtedly talented in music, but the father preferred something more reliable. In 1581, when Galileo turned seventeen, the scales were leaning in the direction of medicine. Vincenzio understood the expenses of instruction would be great, but his son's future would be assured. The place of instruction was chosen to be the University of Pisa, perhaps a bit provincial, but familiar to Vincenzio. He had lived for a long time in Pisa, and Galileo was born there.

The road to becoming a doctor was not easy. Before beginning to study medicine, it was necessary to learn—by heart—Aristotelian philosophy.

---

[1] Joseph-Louis Lagrange, *Mécanique Analytique*, 5th ed., vol. 1 (Paris: Blanchard, 1965), p. 207.

In Galileo's opinion, "it seems that there is not a single phenomenon worth attention that he [Aristotle] would have encountered without considering." At the time, Aristotelian philosophy was taught in an appalling way, namely as a selection of statements considered to be the ultimate truth, devoid of motivation or proof. One could not even talk about disagreeing with Aristotle.

What interested Galileo most of all was what Aristotle wrote about the physics of the world around him, but he did not want to believe every word of the great philosopher blindly. He mastered it by using logic: "Aristotle himself taught me to be satisfied in my mind only when the arguments convinced me, and not just the teacher's authority." He also read other authors, and Archimedes and Euclid were among those who impressed him the most.

Mysteries of Motion

Of everything that takes place in the world around us, Galileo was most interested in motion in its various forms. Bit by bit, he gathered everything the ancients wrote about motion, but regrets, "There is nothing older than motion in nature, but rather little that is significant has been written about it." Questions came to the inquisitive youth at each step. . . .

"In 1583, at the age of about twenty, Galileo found himself in Pisa, where, on his father's advice he applied himself to the study of philosophy and medicine. And one day, being in the cathedral of that city and curious and clever as he was, he decided to observe the motion of a [hanging] chandelier that swerved from the perpendicular—whether the time it took to swing back and forth along long arcs was the same as along medium and short arcs. It seemed to him that the time to travel a long arc might be less because of the greater speed with which, as he saw, the lamp moved along the higher and more sloped sections. And since the lamp moved gradually, he made a gross estimate, as he was wont to say, of how it moved back and forth, using the beating of his own pulse, as well as the tempo of music he had with great profit practiced. And on the basis of these calculations he saw that he was not mistaken in believing that the times were the same. But not satisfied with this, on returning home he thought of doing the following, in order to make certain.

"He attached two lead balls to strings of exactly the same length so that they could swing freely. . ., and displacing them from the vertical by different numbers of degrees, for example one by 30 and the other by 10, he released them at the same instant. With the help of a friend he observed that while one made a certain number of oscillations along long arcs, the other made exactly the same number along small ones.

"In addition he made two similar pendula, but of rather different lengths. He observed that while the short one made a certain number of oscillations, for example, 300 along its longest arcs, in the same time that

Galileo Galilei (seventeenth century engraving), 1564–1642.

the long one always made the same number, say 40, both along its longest arcs and its shortest ones; repeating this several times..., he concluded from this that the time to go back and forth is the same for the same pendulum, the longest or the shortest, and that there are almost no notable differences in this which must be attributed to interference by the air, which resists a faster moving heavy object more than a slowly moving one.

"He also saw that neither different absolute weights, nor different specific gravities of the balls made any manifest change in this—all, provided they hang from strings of equal length from their centers to their points of suspension, keep much the same time to travel along every arc; as long as one does not take a very light material, such as cork, whose motion in air...is more easily resisted and which more quickly comes to rest."

This story is due to Vincenzio Viviani[2] (1622–1703), who in 1639, at the age of seventeen, was at the villa of Arcetri near Florence, where Galileo

---

[2] Letter to Prince Leopoldo de' Medici, 1659, in Galileo Galilei, *Opere*, ed. A. Favaro, vol. 19 (Florence: Barbèra, 1938), pp. 648–649.

found himself after the verdict of the Inquisition. For two years, Evangelista Torricelli (1608–1647) was also there, and the two helped the blind scientist complete his projects. They obtained a series of results under Galileo's influence (famous barometric experiments and research on cycloids). Viviani was apparently especially close to Galileo, who willingly discussed various subjects with him, often recalling the distant past. Afterwards, Viviani had many occasions to retell what he had heard in those days. These tales are not considered sufficiently reliable, and it is not always clear who was the source of the inaccuracies—the narrator or the listener. Viviani's main goal in life was to immortalize his teacher's memory.

Let us return to Viviani's story. He describes Galileo's discovery that a pendulum is isochronous: For a fixed length, the period of its oscillation does not depend on its amplitude. It is instructive to see how Galileo kept time, with music and his pulse (it seems that Cardano was the first to suggest this method). We in the twentieth century, who are accustomed to wristwatches, should not overlook such difficulties. Rather precise clocks were constructed immediately afterwards, based on Galileo's discovery of the pendulum's property (we will have a chance to talk about pendulum clocks later). Incidentally, in laboratory experiments which we will discuss below, Galileo used slowly dripping streams of water to measure time (a variation on water clocks).

Galileo discovered a connection between the length of a pendulum and the frequency of its oscillations: The square of the period is proportional to the length. Viviani wrote that Galileo obtained this result "guided by geometry and by his new science of motion," but no one knows how he could have reached such a theoretical conclusion. Perhaps Galileo observed this relationship experimentally. He apparently did not know that the oscillations of a pendulum are only isochronous for small angles of deflection. For large angles, the period begins to depend on the angle and for 60°, for example, it is noticeably different from the period for small angles. Galileo could have noticed this in the series of experiments Viviani described. The error in Galileo's claim that a mathematical pendulum is isochronous was discovered by Huygens.

His medical studies did not go very well, although Galileo tried to justify his father's hopes and expenditures. In 1585, he returned to Florence without having received a doctor's diploma. There, he continued to study mathematics and physics, first in secret from his father and then with his agreement. Galileo was in contact with scientists, including the marchese Guidobaldo del Monte. Thanks to the latter's support, in 1589 Ferdinando de' Medici, the Grand Duke of Tuscany, appointed Galileo as professor of mathematics at the University of Pisa. Galileo remained in Pisa until moving to Padua in 1592. He considered his eighteen years in Padua the happiest period of his life. From 1610 until the end of his life, he was "Philosopher and First Mathematician of his Highness the Grand Duke of

Tuscany." In both Pisa and Padua, the study of motion was Galileo's main work.

Free Fall

Galileo was above all interested in free fall, one of the most common forms of motion in nature. At the time, one had to begin with what Aristotle said on the matter. "Bodies having a greater degree of heaviness or lightness but in all other respects having the same shape, traverse an equal space more rapidly in the same proportion as the quantities mentioned." Thus, according to Aristotle, the velocity of a falling body is proportional to its weight. A second assertion is that velocity is inversely proportional to "the density of the medium." This assertion led to complications, since in a vacuum, whose "density" is zero, the velocity should have been infinite. As to this, Aristotle declared that a vacuum cannot exist in nature ("nature abhors a vacuum").

Aristotle's first assertion was sometimes disputed, even during the Middle Ages. But Benedetti's criticism was especially convincing. Benedetti was Tartaglia's student and Galileo's contemporary, and Galileo became familiar with his treatise in 1585. The main idea of Benedetti's refutation looks like this. Suppose we have two bodies, one heavy and one light: The heavy one should fall faster. Now combine them. It is natural to assume that the light body slows down the heavy one, and that the velocity at which the combined body falls should be intermediate between the individual velocities. But according to Aristotle, the velocity should be greater than that of each body! Benedetti decided that velocity depends on specific gravity, and even estimates that for lead it is eleven times greater than for wood. Even Galileo believed this for a long time.

Galileo began studying free fall in Pisa. Here is what Viviani writes: "...Galileo completely gave himself up to reflection, and to the great embarassment of all philosophers, he was persuaded, by means of experiments, solid proofs, and arguments, of the falsity of very many of Aristotle's conclusions about motion that up to that time had been considered perfectly obvious and unquestionable. These include, among others, that two bodies of the same material but different weight, moving in the same medium, do not move with speeds proportional to their weights as Aristotle proposed, but with the same speed. He proved this by repeated experiments from the top of the Tower of Pisa, in the presence of other lecturers and philosophers and the entire scholarly fraternity."[3] To this time, Galileo is often drawn throwing balls from the Tower of Pisa. This legend has acquired many spicy details (for instance, the bartender who started the rumor that Professor Galilei would jump from the tower). Note that so far only bodies of the same substance are being discussed.

---

[3] Ibid., p. 606 (letter of 1654).

Galileo studied Benedetti's observation that the velocity of free fall increases according to the body's motion and decided to find a mathematically precise description of this change. Here we should say that from the start Galileo saw his problem as how to quantify Aristotle's physics: "Philosophy is written in that great book, which ever lies before our eyes (I mean the universe); but we cannot understand it if we do not first learn the language and grasp the symbols in which it is written. The book is written in the mathematical language and the symbols are triangles, circles, and other geometrical figures...."[4] However, it soon became clear that quantification requires a systematic review of all the facts.

How, then, to find the law by which the velocity of free fall changes? An experiment was only the beginning of scientific research. Aristotle and his followers considered experimentation unnecessary and worthless, for both establishing and verifying the truth. Galileo could have tried to conduct a series of experiments on bodies in free fall, carry out careful measurements, and search for a law that would explain them. This is the way Galileo's contemporary, Kepler, working with Tycho Brahe's many observations, discovered that the planets move along ellipses. But Galileo chose a different route. He decided to guess the law first from general considerations, and then to verify it experimentally. No one had done this before, but gradually this way of doing research has become one of the leading methods of establishing truth in science.

Now let us see how Galileo tried to guess the law. He decided that nature "tries to take the simplest and easiest way in all its adaptations," which means that the law by which the velocity grows must be "in the simplest and most universally clear form." But since the velocity grows according to the distance traveled, what could be simpler than assuming it is proportional to the distance: $v = cs$, where c is a constant. This was wrong from the start; after all, it would imply that free fall begins with zero velocity, while the velocity is apparently large from the very beginning. But here is an argument that convinced him there is no contradiction: "...is it not true that if a block be allowed to fall upon a stake from a height of four cubits and drives it into the earth, say, four finger-breadths, that coming from a height of two cubits it will drive the stake a much less distance, and from the height of one cubit a still less distance; and finally if the block be lifted only one finger-breadth how much more will it accomplish than if merely laid on top of the stake without percussion? Certainly very little. If it be lifted only the thickness of a leaf, the effect will be altogether imperceptible. And since the effect of the blow depends upon the velocity of this striking body, can anyone doubt the motion is very slow and the speed more than small whenever the effect [of the blow] is imperceptible?"[5]

---

[4] Galileo, *Opere*, vol. 4, p. 171.

[5] Galileo, *Discorsi e Dimostrazioni Matematiche, Intorno à Due Nuove Scienze (Dialogues Concerning Two New Sciences)*, trans. Henry Crew and Alfonso de Salvio (New York: Dover, 1954), pp. 163–164.

For a long time, Galileo studied the various consequences of this assumption and unexpectedly discovered that...*according to such a law motion cannot take place at all!* Let us also try to see why this is so. The proportionality coefficient depends on the choice of the unit of time. For simplicity, we will assume that $c = 1$, that distance is measured in meters (m), and time in seconds (sec). Then at all moments of time, $v = s$.

Consider a point A one meter from the origin O. Let us estimate at what time the body reaches this point after it begins to move. At A the velocity equals 1 meter per second (m/sec). Take the point A_1 halfway between O and A. At each point in the interval A_1A the instantaneous velocity will be less than 1 m/sec, and since this interval has length 1/2 m, more than 1/2 sec is required to traverse it. Now take the point A_2 halfway between O and A_1. On A_2A_1 the instantaneous velocity will be less than 1/2 m/sec (all its points are less than 1/2 m from O), and since its length is 1/4 m it too requires more than 1/2 sec. Of course, you have already guessed the rest of the argument: A_3 is the midpoint of OA_2, the length of A_3A_2 is 1/8 m, the velocity is less than 1/4 m/sec, and again more than 1/2 sec is required, etc. The division process can be continued endlessly, and we can choose any number of intervals, each requiring more than 1/2 sec, without reaching O. This means that a body leaving O cannot arrive at A!

We assumed that A is at a distance of 1 m from O. But it can be shown analogously that a body leaving O can reach no point whatsoever. This remarkable argument was the beginning of classical mechanics!

However, Galileo himself published an unconvincing argument. He tried to reach a contradiction by saying that since velocity is proportional to distance, all intervals beginning at the origin must be traversed in the same time, which is impossible. Either Galileo was not yet used to working with instantaneous velocity, or he originally had some other argument that he could not reconstruct when he wrote down these results, in his old age, after a long interruption (we will see why this happened). He left more than a few claims that were either without motivation or supported by doubtful arguments.

Well, Galileo had every reason to be offended by the perfidy of nature, which did not choose the simplest path, but he did not lose his belief in nature's reasonableness. He considered a no-less-simple assumption, that the velocity grows proportionally with time: $v = at$. He called this naturally accelerated motion, but the term "uniformly accelerated motion" has survived. Galileo considered the graph of the velocity on the time interval from O to t and remarked that if we take moments of time t_1, t_2 equidistant from $t/2$, then the velocity at t_2 is greater than $at/2$ by the same amount that it is less than $at/2$ at t_1. From this he concluded that at the midpoint the velocity equals $at/2$, and the distance traveled equals $(at/2)t = at^2/2$ (not too rigorous an argument!). This means that *if we consider equally spaced moments of time $t = 1, 2, 3, 4, \ldots$, then the corresponding distances traveled from the origin will be proportional to the squares $t^2 = 1, 4, 9, 16,$*

..., *and the distances traveled between adjacent times proportional to the odd integers* 1, 3, 5, 7, ...

Again, let us follow Galileo's logic. First he separated the questions "how" and "why." For Aristotle's followers, the answer to the first question had to be an immediate consequence of the answer to the second. But Galileo, soberly evaluating his chances, did not investigate the origin of accelerated, free-fall motion in nature but instead only tried to describe the law by which it occurs. The main thing was to search for a simple, general principle from which this law can be deduced. He sought "a completely unquestioned principle, that can be taken as an axiom." Galileo's statements in a letter to Paolo Sarpi (in the autumn of 1604) can be interpreted to say that he already knew the law by which the distance traveled changes during free fall but was dissatisfied because he could not deduce it from an apparently unquestioned principle: "A body experiencing natural motion increases its velocity in the same proportion as its distance from the initial point."

Here it was important to choose a fundamental independent variable, relative to which the change in all quantities characterizing motion could be considered. It is very natural to begin by choosing the distance traveled as this variable; after all, an observer sees how the velocity grows as the distance grows. We have already said that the measurement of time did not yet play a significant role in people's lives and that precise clocks were not available. We do not always take into account how gradually the sensation of constantly passing time took root in human psychology. Galileo showed great flexibility in reorienting himself comparatively quickly from distance to time. During 1609–1610 he discovered the true principle that free fall is uniformly accelerated (with respect to time!).

We must not overemphasize the final form of Galileo's ideas of velocity and acceleration. The idea of an instantaneous, continuously changing velocity is not easy to sense, and it won acceptance slowly. It was hard to convince oneself that rejecting abrupt changes in velocity did not lead to the contradictions with which arguments about continuous procedures overflowed. It is difficult for us today to judge Galileo's courage in working with a varying velocity so decisively. Such masters of analytical arguments as Cavalieri, Mersenne, and Descartes did not believe him. Descartes categorically rejected motion with zero initial velocity, in which a body "passes through all degrees of slowness." The process of calculating distances under a varying velocity was even more complicated, requiring integration. Galileo possessed only a variant of integration, similar to Archimedes' method or to Cavalieri's "indivisibles." In this case he adopts an artifice, passing to the average velocity in a not well-grounded way and then using the usual formula for uniform motion. Not only the new mechanics but also the new mathematical analysis found its origins in the discovery of the law of free fall. Since Galileo restricted himself to the case of constant acceleration, the concept of acceleration in general was not needed. The acceleration of free fall as a universal constant does not appear in Galileo's work.

When it comes to the role of force in nonuniform motion, Galileo's statements lack complete clarity. He rejected Aristotle's principle that the velocity is proportional to the acting force, because when there is no force, uniform rectilinear motion is maintained. The law of inertia (Newton's first law) carries Galileo's name. Galileo constantly turns to the example of a projectile, which would fly along a line if not for the earth's attraction. He wrote that "the degree of velocity displayed by the body inexorably lies in its own nature, at the same time as the reasons for its acceleration or deceleration are external," "...horizontal motion is eternal, for if it is uniform then it does not weaken for any reason, does not slow down, and is not extinguished." Galileo, in his *Letter to Francesco Ingoli*, poetically describes various phenomena on board a ship, moving uniformly in a straight line, that do not reveal this motion: Drops of water fall exactly into the mouth of a jar placed below, a stone falls straight down from the mast, steam rises straight up, butterflies fly with the same speed in all directions, etc. We have the sense that Galileo confidently supported the principle of inertia in "terrestrial" mechanics, but not in celestial mechanics (more about this later).

Newton ascribed to Galileo not only the first law of mechanics but also the second, although this was an overstatement: Galileo did not make a clear connection between force and acceleration (when they are different from zero). As far as free fall is concerned, Galileo thoroughly answered "how," but not "why."

Motion Along an Inclined Plane

Galileo considered his most fundamental conclusion to be that during consecutive equal time intervals, a falling body travels distances that are proportional to consecutive odd numbers. He wanted to verify this, but how? He could not continue to throw balls from the Tower of Pisa, since he was already living in Padua. In the laboratory, free fall takes place very quickly. But Galileo found a clever way out: He replaced free fall with the slower movement of a body along an inclined plane. He noticed that assuming free fall is uniformly accelerated implies that the movement of a point mass along an inclined plane is also uniformly accelerated. This is essentially today's common argument of resolving forces, showing that a point mass slides along an inclined plane with constant acceleration $g \sin \alpha$, where α is the angle of inclination to the horizontal and g is the acceleration of free fall. Galileo's reasoning was more awkward: He did not introduce the acceleration of free fall but instead manipulated a large number of proportions, as was then common. He drew a whole series of consequences from the uniform acceleration of a point on an inclined plane which could be conveniently verified in the laboratory (because when the angle is small, it takes a long time for the point to slide down the plane). A key assertion is that if inclined planes have the same height, then the sliding times are related as the distances traveled (why?).

Motion along an inclined plane was a question of independent interest for Galileo, and he made many observations. For example, if points move along chords AE_i and BF_j of a circle, where AB is vertical and a diameter,

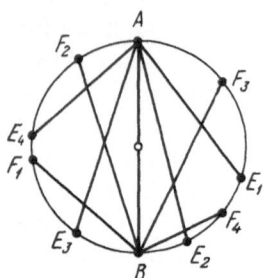

then all sliding times equal the time for free fall along AB (prove it!). A rather complicated argument leads Galileo to the proof that if A, B, and C are successive points along a circle, then a point slides faster along the polygonal line ABC than along the chord AC. This is associated with a well-known error of Galileo: He assumed that a point moves most quickly along a quarter of a circle, but this is true instead for an arc of a cycloid.

Projectile Motion

Galileo called the motion of a projectile *forced* (as opposed to free fall). Aristotle assumed that a body thrown at an angle to the horizontal first moves along an inclined line, then along a circular arc, and finally along a vertical line. Tartaglia may have been the first to claim that the trajectory of a projectile "does not have a single part that is perfectly straight."

Galileo constructed a theory of forced motion immediately after his theory of free fall. His route was the same: The theory (a model of the phenomenon) preceded experiments. Galileo's guess was brilliantly simple: The motion of a projectile launched at an angle to the horizontal is made up of the uniform rectilinear motion that would occur if it were not for the force of gravity, together with free fall. In the end the body moves along a parabola. Let us note that this argument essentially uses the law of inertia—Galileo's law.

In considering complex motion, Galileo had a brilliant predecessor who served as his model: "I assume nothing except the definition of the motion I wish to treat of and whose properties I wish to demonstrate, imitating in this Archimedes in his *Spiral Lines*, where he, having explained what he means by motion made in the spiral that is compounded from two uniform motions, one straight and the other circular, goes on immediately to demonstrate its properties."[6] This refers to the spiral of Archimedes, described by a point moving along the radius of a rotating circle (e.g., a fly walking towards the center of a phonograph record).

Using the properties of a parabola, Galileo constructed a "table for gunners, having important practical significance." It was not without reason that Padua belonged to the Venetian republic, and Galileo was in constant contact with the Venetian arsenal. Many of his claims, reached theoretically, could be verified experimentally. He proved Tartaglia's assertion that a shot travels farthest for a 45° angle, and showed that for angles adding up to 90° shots fired with the same velocity travel the same distance.

Galileo and Kepler

Galileo's discoveries must have startled his contemporaries. The conic sections—ellipses, parabolas, and hyperbolas, the acme of Greek geometry—seemed to be the fruit of mathematical fantasy, unrelated to reality. And here Galileo showed that parabolas inevitably arise in a perfectly "worldly" situation. (Even in the nineteenth century, Laplace presented an application of the conic sections as a most unexpected use of pure mathematics.) It is remarkable that literally at the same time, the conic sections arose in a completely different problem and in a no less surprising way. In 1604–1605, Johannes Kepler (1571–1630) discovered that Mars moves along an ellipse, with the sun at one focus (within ten years Kepler extended this to all the planets). This is an important coincidence and for us the two discoveries go hand in hand, but it is likely that before Newton no one seriously put these results together. Moreover, Galileo did not accept Kepler's law and did not communicate his own discovery to Kepler, in spite of their regular correspondence (only published after Kepler's death).

Galileo and Kepler corresponded for many years. Kepler was one of the scientists closest in spirit to Galileo. First of all it was essential that Kepler unreservedly accepted the Copernican system. As early as 1597, Galileo, in connection with receiving Kepler's book *Mysterium Cosmographicum* (*Cosmographic Mystery*), shared with Kepler his secret desire to publish his arguments supporting the Copernican system. He wrote: "...I do not publish them, because I am deterred by the fate of our teacher Copernicus, who, though he had won immortal fame with a few, was ridiculed and condemned by countless people (for there are so many fools). I would dare to publish my speculations if there were more people like you." In response, Kepler sent a passionate appeal: "Be of good cheer, Galileo, and come out publicly." He proposed joining forces: "If I judge correctly, there are only

---

[6] From a 1639 letter to Giovanni Battista Baliani, trans. Stillman Drake, *Galileo at Work* (Chicago: Univ. of Chicago Press, 1978), pp. 395–396.—*Transl.*

a few among the distinguished mathematicians of Europe who would part company with us...." And a book can also be printed in Germany, not just in Italy. The problem was seen differently in far off Prague than in Italy, where for the sixth year Giordano Bruno[7] awaited his fate in prison.

The path Kepler took towards his discovery is very instructive. Kepler had two faces as a scientist. On the one hand, he was a great visionary, attempting to understand the greatest mysteries of the universe. He was certain that the greatest mystery he discovered was that there are six planets because there are five regular polyhedra! "I have never succeeded in finding words to express my delight at this discovery." Kepler proposed six spheres alternating with the regular polyhedra, so that for each sphere we have one inscribed and one circumscribed polyhedron. He placed the spheres in correspondence with the successive planets. There is a special scheme hidden in the order of the polyhedra (the cube corresponds to Saturn, the tetrahedron to Jupiter, etc.). Kepler compares the ratios of the radii of the spheres with the known relative dimensions of the orbits and, strangely enough, finds only a small discrepancy (except for Mercury). These arguments, published in *Mysterium Cosmographicum*, met with favor from many people and with no objection from Galileo, and Tycho Brahe, the "king of astronomers," invited Kepler to collaborate with him.

Another side of Kepler's scientific life, so different from this one, is associated with this invitation. He scrupulously worked through Tycho's many observations, which were unprecedentedly accurate for observations made without a telescope (their accuracy is estimated at $\pm 25''$). He needed to reexamine the planetary orbits, using Tycho's data. Tycho (whom Kepler called the "phoenix of astronomy") evidently expected a corroboration of his own compromise theory, in which the sun moves around the earth and the remaining planets move around the sun. But Kepler carried out his computations in the Copernican system.

Since Copernicus, like Ptolemy, formed the planetary orbits from circles, his system retained the use of epicycles. Kepler wanted to simplify the system (his summary work, appearing in 1618–1621, was called *Epitome Astronomiae Copernicanae (Epitome of the Copernican System)*). Surprisingly, Earth's orbit differs only slightly from a circle, but the sun is somewhat off-center. Copernicus knew all this, but Kepler made the size of this displacement precise. He carefully studied the nonuniformity of the earth's motion in its orbit and for a long time searched for a law to describe it. He tried an inverse proportional dependence on the distance from the sun as well as various other possibilities, but did not yet discover the law of areas (Kepler's second law). Then he computed Mars' orbit and compared it to different curves. Kepler's attitude towards observational data was startlingly sober and confident. Once he rejected a hypothesis

---

[7] a poet and philosopher who dreamt of other words—*Transl.*

because he discovered a disrepancy of 8' from Tycho's data (such a discrepancy is almost invisible to the naked eye). Einstein said, "He first had to recognize that even the most lucidly logical mathematical theory was of itself no guarantee of truth, becoming meaningless unless it was checked against the most exacting observations in natural science."[8] Kepler looked over different sorts of ovals, finally discovering that an ellipse with the sun at one focus fit the data. "Groping incessantly everywhere in the surrounding gloom, I emerged, finally, into the clear light of truth." Isn't Kepler's way very different from Galileo's? To a great extent, Galileo proceeded from general principles and qualitative results. In his declining years, Galileo recalled: "I have always esteemed Kepler a free mind (perhaps even too much) and acute, but my way of philosophizing is utterly different from his; it may be that writing on the same matters, and strictly as concerns celestial motions, we have both hit on the same idea, but only in a very few instances, whereby we assigned of some true effect the same true reason; but this will be found to be the case in less than one in a hundred of my thoughts."[9]

Galileo believed that uniform circular motion rules the universe. He believed neither in elliptical orbits nor in the nonuniform motion of the planets in their orbits, and did not take data of observational and computational astronomy into account.

Kepler was the first to consider the mutual attraction of bodies and associate it with motion; he even conjectured how the interaction decreases with distance (as $1/r$, which is false). He explained the tides by lunar attraction. All this was totally unacceptable to Galileo, who rejected forces acting from afar and, in particular, attempts to explain terrestrial phenomena by the influence of heavenly bodies. This related particularly to the tides, which Galileo erroneously assumed to be an important proof of the earth's motion. Galileo identified such an explanation with astrology, in which events in human life are explained by the influence of the planets. "Among the great people discussing this startling phenomenon of nature, the one who surprises me most is Kepler, who, possessing a free and sharp mind and being quite familiar with the motions ascribed to the Earth, admits a fundamental power of the Moon on water, hidden properties, and such similar childishness." Kepler turned out to be right, but the real arguments appeared later.

We must keep in mind that Kepler's arguments on mutual attraction are

---

[8] Albert Einstein, *Out of My Later Years* (New York: Philosophical Library, 1950), p. 226.
[9] From a 1634 letter to Fulgenzio Micanzio. This translation, together with other comments on Galileo's relation to Kepler, appears in Giorgio de Santillana's translation of *Dialogo...Sopra i Due Massimi Sistemi del Mondo* (*Dialogue on the Two Great World Systems*) (Chicago: Univ. of Chicago Press, 1953), pp. 349–350.—*Transl.*

often confused. In one respect, he seriously lagged behind Galileo: He assumed, following Aristotle, that velocity is proportional to force.

Terrestrial and Celestial Mechanics

Towards 1610, Galileo obtained results in mechanics towards which he had been working for twenty years. He began work on a comprehensive treatise, but unexpected events diverted him from this for more than twenty years! Galileo built a telescope, and at the beginning of 1610 he discovered the moons of Jupiter. For that entire year, astronomical discoveries followed one after the other. Galileo thought he had decisive proof supporting the Copernican system. He completely devoted the following twenty-three years of his life to confirming it, until in 1633 the verdict of the Inquisition interrupted this activity. All these years Galileo thought about mechanics, since this was required by his work *Dialogo...* *Sopra i Due Massimi Sistemi del Mondo* (*Dialogue on the Two Great World Systems*). His new philosophy even temporarily contradicted results about "terrestrial" motion. Thus he did not find a place in the universe, "where all parts are in the most excellent order," for rectilinear motion, which under these conditions seemed to him "unnecessary and unnatural." The reason for this was that motion along a line cannot be periodic and the state of the universe must always be changing. He left room for rectilinear motion only in unstable situations, while in nature circular motion must rule. Galileo considered the law of inertia he discovered for "local motion" to be valid only near Earth.

Galileo also considered the parabolic law of motion for projectiles to be approximate. He assumed that the trajectory would in fact end at the center of the earth. Because of this he made strange pronouncements about the motion of a projectile having to follow a circular arc or spiral, even after he discovered the trajectory was parabolic. Fermat objected to this, communicated through Carcavi (1637). In reply, Galileo called his statement "poetic fiction" and promised to publish his claim that the trajectory was parabolic, but wrote in conclusion: "But in the motion I defined it is found that all the properties that I demonstrate are verified... in that when we make experiments upon the earth and at heights and distances practicable to us, no sensible difference is discovered, though a sensible, great, and immense difference would be made by approaching to and closely nearing the center."[10] The approximate nature of the parabolic trajectory was clarified by Newton, but Galileo's expectations were unwarranted.[11]

---

[10] For an English translation of this letter, see Drake, *Galileo at Work*, pp. 376–381.—*Transl.*

[11] Since Galileo delayed publication for a long time, the first mention of the parabolic trajectory appeared in 1632 in Cavalieri's *Lo Specchio Ustorio* (*The*

The main question about motion that interested Galileo all these years was linked to the standard objection of the opponents of the earth's motion: Why don't objects fly off a moving earth? Galileo had no doubt that the force of gravity was responsible for this, but how could he give a motivated explanation? Let a body move on a sphere of radius R with velocity v. This is how Galileo began his arguments. Let us first fix a starting point. If it were not for the force of gravity the body would continue in rectilinear motion, with velocity v along a tangent. To guarantee motion on the sphere (to hold onto the body), we must add motion in the direction of the center. Galileo was used to combining motions! What was left to do? Just to note that (by the Pythagorean theorem) for the second motion the distance traveled is $s(t) = \sqrt{R^2 + v^2 t^2}$ $- R$, and if the time t is small then this is almost the same as $s^*(t) = v^2 t^2 / 2R$ [since $(s - s^*)/t^3 \to 0$ as $t \to 0$]. Now we cannot fail to recognize Galileo's formula for distance under uniformly accelerated motion, with acceleration $a = v^2/R$. So if $g > a$, then the body will stay on the surface of the sphere. But Galileo did not carry out the second half of this reasoning, giving instead a very confused motivational argument. The formula for centripetal acceleration, along the path Galileo outlined, was obtained by Huygens in 1659.

Discorsi (Two New Sciences)

Finding himself in exile in Siena in 1633, several weeks after the verdict of the Inquisition and his renunciation, Galileo thought about the results he obtained in mechanics long ago and decided to publish them immediately. He continued work in Arcetri and Florence, without regard to his enforced isolation, worsening health, and progressing blindness. He wrote, "Even though I am silent, I do not live my life completely idly." His book *Discorsi e Dimostrazioni Matematiche, Intorno à Due Nuove Scienze Attenenti alla Mecanica e i Movimenti Locali* (*Discourses and Mathematical Proofs, Concerning Two New Sciences Pertaining to Mechanics and Local Motions*) was completed in 1636, was sent across the border with great precautions (it was not clear how the Inquisition regarded the book), and came out in Holland in July of 1638. As with his previous book *Massimi Sistemi*, which was the reason for his persecution, Galileo wrote *Discorsi* in dialogue form.[12] The dialogues take place over the course of six days and

---

Burning Glass), which very clearly took from Galileo the idea of adding rectilinear motions and the principle of inertia. Galileo was offended by the absence of the necessary references, and spoke of his discovery that the trajectory is parabolic as the main goal of forty years' work. Cavalieri's apologies quickly satisfied Galileo.

[12] There is some confusion over the English titles of these books since both have been translated as *Dialogues...*, as noted in earlier references. To avoid confusion, we will use the abbreviated Italian titles.—*Transl.*

DISCORSI
E
DIMOSTRAZIONI
MATEMATICHE,
intorno à due nuoue scienze

Attenenti alla
MECANICA & i MOVIMENTI LOCALI,

del Signor
GALILEO GALILEI LINCEO,
Filofofo e Matematico primario del Sereniffimo
Grand Duca di Tofcana.

Con vna Appendice del centro di grauità d'alcuni Solidi.

IN LEIDA,
Appreffo gli Elfevirii. M. D. C. XXXVIII.

Title page of *Discorsi*.

involve the same heroes: Salviati (taking the author's point of view), Sagredo, and Simplicio (who is on Aristotle's side; his name means "simpleton"). On the third and fourth days they read an academic treatise on motion (by Galileo), *De Motu Locali* (*On Local Motion*), and discuss it in detail. Incidentally, in the title of the book "mechanics" and "motion" are separate, since at that time mechanics meant only statics and strength of materials. The form chosen by Galileo for the discussion allowed many to realize how he approached his own discoveries.

The aged Galileo strove to realize the ideas he had laid aside for so long. But much was already beyond his power, and he needed assistants. He entrusted his son Vincenzio with constructing a clock based on the discoveries in his youth about the pendulum, but he did not see his ideas brought to fruition. The Inquisition had limited Galileo's contacts with the outside world. After the completion of *Discorsi* at the villa at Arcetri, which Galileo called his prison, much-wanted guests began to appear. These were his old friend and trusted student Benedetto Castelli, and Cavalieri. Also Viviani and Torricelli did not abandon their teacher, but for some time had been helping him complete his work and continue his research.

Torricelli computed the velocity vector for a projectile launched at an angle, using the addition rule for velocities, and since the velocity is directed along a tangent he obtained an elegant way to extend the tangent to a parabola. The era of differential and integral calculus had arrived and problems about extending tangents to curves came into the foreground of mathematics. Various methods were worked out to solve them. One became the kinematic method, in which the curve was represented as the trajectory of a complex motion and the tangent was found by adding velocities, as Torricelli first did for parabolas. The French mathematician Gilles Persone de Roberval (1602–1675) worked miracles with this method. "Mechanical" curves, obtained as the trajectories of various motions, firmly came into use in mathematical analysis. It is worth recalling that Galileo consciously limited himself to motions that arose in nature: "For anyone may invent an arbitrary type of motion and discuss its properties; thus, for instance, some have imagined helices and conchoids as described by certain motions which are not met with in nature, and have very commendably established the properties which these curves possess... but we have decided to consider the phenomena of bodies...such as actually occurs in nature."[13] The value of a general view of motion was demonstrated by Newton.

Discorsi determined the development of mechanics for a long time. It was on the desks of both Huygens and Newton, Galileo's great successors. It is difficult to imagine how much the development of mechanics would have been delayed if these unfortunate events had not occurred and if Galileo had not written down his great discoveries.

Mathematical Appendix

The history of the discovery of the law of free fall has another side: This is the history not only of a discovery that was carried through but also of one that was neglected. Once Galileo understood that motion could not take place according to $v(t) = c \cdot s(t)$, he lost interest in this law. He was only interested in motions that occur in nature! Soon the Scottish Lord Napier became interested in motion that follows an analogous law.

Napier considered rectilinear motion taking place according to the law $v(t) = L(t)$, where $v(t)$ is the instantaneous velocity at time t and $L(t)$ is not

---

[13] *Discorsi*, p. 160.

the distance traveled, but rather the distance between the moving point at time t and a fixed point O on the line. The case Galileo considered corresponds to the case when the moving point is at O at the initial time $t = 0$, i.e., $L(0) = 0$, $L(t) = s(t)$. For Napier, $L(0) > 0$, and $L(t) = L(0) + s(t)$.

It turns out that for $L(0) > 0$, such motion can occur in principle and possesses remarkable mathematical properties (although it does not "occur in nature"). Let's examine this. First of all, if the initial distance $L(0)$ is multiplied by c, then at each moment of time $L(t)$ and $v(t)$ are also multiplied by c. Strictly speaking, this requires proof! But it is clear that after multiplying L and v by a constant, the law $v(t) = L(t)$ still holds. Next, we restrict ourselves to the case $L(0) = 1$. Then, as we will show,

$$L(t_1 + t_2) = L(t_1)L(t_2).$$

It is convenient to take t_1 as a new time origin. By taking $\dot{c} = L(t_1)$, we see that at the new time t_2 (the old $t_1 + t_2$) the distance to O should be $L(t_1)$ times greater than at the old t_2. But this means that $L(t_1 + t_2) = L(t_1)L(t_2)$. This is how the exponential function first appeared in science.

We have $L(t) = e^t$, where $e = L(1)$, i.e., the distance from 0 at $t = 1$. Using this and $v = L$, it is not difficult to show that $e > 2$ (prove it!). In fact, $e = 2.71828\ldots$ and became known as Napier's number. By considering motion according to the law $v(t) = kL(t)$, we can obtain exponential functions with other bases.

For any positive number a, we will call the time t for which $L(t) = a$ the (natural) logarithm of a (denoted by $\ln a$).[14] By the above, $\ln ab = \ln a + \ln b$. For twenty years Napier composed tables of logarithms, and in 1614 published *Mirifici Logarithmorum Canonis Descriptio* (*A Description of the Wonderful Law of Logarithms*), whose foreword contained an apology for the inevitable errors and finished with the words, "Nothing is perfect at first."

Napier's discovery is remarkable not only because he constructed logarithm tables but also because he showed that new functions can appear through the study of motion. Beginning with the work of Galileo and Napier, mechanics became a constant source of new functions and curves for mathematics.

II. The Medicean Stars

In November 1979, the Vatican decided to rehabilitate Galileo, who had been sentenced by the Inquisition in 1633. Then he had been acknowledged to be "vehemently suspected of heresy" because "he held and

---

[14] Napier's reasoning was not quite the same as this and Napier's logarithms are different from natural logarithms.

defended as probable" the opinion that "the Sun is the center of the world and does not move from east to west, and that the Earth moves and is not the center of the world."[1] At a session of the Vatican Academy of Sciences in November 1979, devoted to Einstein's centenary, Pope John Paul II noted that Galileo "suffered greatly—we cannot conceal this now—from oppression on the part of the Church," but qualifying Galileo's repentance as "divine illumination in the mind of the scientist," he asserted that Galileo's tragedy emphasizes the "harmony of faith and knowledge, religion and science." In October 1980, it was announced that the pope had ordered a supplementary investigation into the circumstances of the proceedings against Galileo. Conversations about Galileo's acquittal had already taken place at the second Vatican Council (1962–1965). They had wanted to time his acquittal to the 400th anniversary, in 1964, of his birth, but evidently did not succeed, since the question turned out to be not uncontroversial. Incidentally, Galileo's works (together with those of Copernicus and Kepler) had been taken off the *Index Librorum Prohibitorum* (*Index of Forbidden Books*) as early as 1835. The judgment against Galileo, and his renunciation, have not ceased to disturb people, often far from science, for three and a half centuries. The attention given by literature to this question is typical (it suffices to recall Berthold Brecht's play, *Galileo*). The problem of Galileo is alive even today, notwithstanding his recent "rehabilitation."

At the dawn of the seventeenth century, the question of a world system was not a simple one. In the fourth century B.C., Aristotle claimed that the seven visible heavenly bodies revolve uniformly around the earth, on crystal spheres to which they are attached, and that the fixed stars occupy an eighth sphere. Astrologers classified the planets as follows: two luminaries, the moon and sun; two harmful planets, Mars and Saturn; two favorable ones, Jupiter and Venus; and one neutral one, Mercury.

Aristotle's rules, and especially those of his followers, did not explain any deviations from his scheme, e.g., the surprising "retrograde" motion, in which a planet appears to reverse direction. Contradictions to precise observational data gradually accumulated. In the second century A.D., Ptolemy constructed a system that took observations into account as much as possible. He assumed that the planets move along auxiliary curves (epicycles) whose centers (deferents) in turn revolve around the earth. The desire to account for new data led to an increasingly complicated system. We must acknowledge the persistence and cleverness by which scientists succeeded in saving this system.

Nicolaus Copernicus (1473–1543) proposed a completely unexpected

---

[1] Complete English translations of Galileo's sentence and renunciation appear in de Santillana, *The Crime of Galileo* (Chicago: Univ. of Chicago Press, 1955), pp. 306 ff.—*Transl.*

route. His carefully worked-out scheme, consistent with observations, contains all the fundamental aspects of today's view of the solar system: The planets, including the earth, revolve around the sun; the earth executes a daily motion; and the moon revolves around the earth. Such an approach simplified everything incredibly, although obscure aspects remained as a result of making the system agree with observational data. In the opinion of Copernicus, the planets move with nearly uniform circular motion (as for Aristotle) but undoubtedly deviate from it. Epicycles were still required, but they played a less essential role than for Ptolemy. The epicycles disappeared only with Kepler, who discovered that the orbits are elliptical. The Copernican system was not a purely descriptive theory based on qualitative phenomena. It contained many computations, e.g., the distance to the sun, the periods of revolution, etc. Only such a theory could compete with Ptolemy's, which completely accounted for the data.

The Pythagoreans had already indicated the possibility of the earth's motion. Therefore, the Church called the teaching that the earth moves Pythagorean. They preferred not to use Copernicus' name for this teaching, for the following reason. His book *De Revolutionibus Orbium Coelestium* (*On the Revolutions of the Heavenly Spheres*), which appeared within a year of his death, was preceded by a foreword that Copernicus himself may not have written. The foreword calls his system a convenient mathematical scheme for astronomical computations and no more, and the motions it considers are called imaginary. This means that "real" motions are not discussed in the book. That is not a mathematician's job! This question must be solved by philosophers and theologians, in accordance with holy writ. The book was dedicated to Pope Paul III. The Church arranged this compromise, and the book was not declared heretical. Mathematicians were allowed to use imaginary schemes in their computations. This included even the Jesuit astronomers, who in particular used Copernicus' tables for the calculations that were necessary to reform the calendar.

The assertion that the earth is stationary and the sun moves had to remain immutable. The Church was not as uncompromising regarding the remaining planets (they are not mentioned in the Scriptures). Tycho's system, in which the sun moves around the earth but the remaining planets move around the sun, was tolerated. Tycho essentially gave up the crystal spheres, asserting that comets do not belong to the "world under the moon" but fly in from the outside (Galileo, by the way, took a different view).

Thus, the Copernican system, a convenient mathematical fiction but a Pythagorean teaching, was heresy. Here lay the boundary. Galileo was not ready to agree to this compromise: "Copernicus, in my view, should not soften, for the motion of the Earth and the immobility of the Sun form the essence and general foundation of his teaching. Therefore they must either condemn it completely or else leave it as it is!" Galileo insisted that the earth's motion is not imaginary, but real.

Galileo's path leading to the decisive struggle for a heliocentric world system was not a simple one. He believed in the Copernican system early on, but for a long time did not decide to publish his arguments in its support (his 1597 letter to Kepler tells us this). The seventeenth century began with the burning of Giordano Bruno. By 1610, Galileo had approached the peak of his scientific activity: He completed his twenty-year-long study of natural motion (free fall and projectile motion) brilliantly. He was beginning work on his great discoveries, and unexpectedly left them for an indefinite period. What happened? Events took place in Galileo's scientific life that forced this completely practical man to put off publishing the discoveries to which he had devoted his youth. Galileo decided that he now had decisive arguments to support the Copernican system and that from then on his life would be wholly given over to propagandizing these ideas. Let us recall these important arguments.

New Glasses

In talking about the lives of great scientists, we rarely pay attention to everyday matters. One of the reasons for Galileo's move from Pisa to Padua was to increase his salary. His material situation became more secure. His initial salary of 180 florins increased, although slowly; he received additional income from the young aristocrats with whom he worked privately and who often lived in his home. But paying his sisters' dowries was a heavy burden that lay on Galileo's shoulders, even as his own family was growing and required greater resources. In 1609 Galileo was anxious about the drawn-out negotiations over increasing his salary. Some sort of invention, with an unquestioned practical application, would open the purse strings of the stingy and practical Venetian *signori*. Galileo was no stranger to technical problems. His home had an excellent workshop, he had recently designed convenient proportional dividers (a "geometric and military compass"), and had himself seen to their manufacture and distribution. He could have thought of artillery tables, based on the parabolic trajectory of a projectile in flight. But he unexpectedly discovered a completely different idea.

In 1608, optical tubes, sometimes called "new glasses," appeared in Holland that allowed people to look at distant objects. Leonardo da Vinci had spoken of glasses through which one could see the moon enlarged, and Roger Bacon of glasses that made a man as big as a mountain. Lippershey and Andriansen, two specialists in optics, contended for the honor of having invented them. Towards the beginning of 1609, such a tube could be bought in Holland for a few *soldi*. By midyear, the tubes had appeared in Paris. Henri IV spoke pessimistically about the innovation, explaining that at the moment he was more in need of glasses that magnified nearby objects rather than distant ones. Then some foreigner tried to sell an optical tube to the Republic of Venice, without going into details about its origins. The first tubes were very imperfect, and Paolo Sarpi, Galileo's friend, expressed a negative opinion about the possibility of using them "in war, on land, and at sea." Galileo heard about the tubes when he was in Venice.

"Upon hearing this news I returned to Padua, where I then resided, and set myself to thinking about the problem. The first night after my return I solved it, and on the following day I constructed the instrument and sent word of this to those same friends at Venice with whom I had discussed the matter the previous day. Immediately afterwards I applied myself to the construction of another and better one, which six days later I took to Venice...."[2] Elsewhere he describes the situation more triumphantly: "...sparing neither labor nor expense, I succeeded in constructing for myself so excellent an instrument that objects seen by means of it appeared nearly one thousand times larger and over thirty times closer than when regarded with our natural vision. It would be superfluous to enumerate the number and importance of the advantages of such an instrument at sea as well as on land."[3]

In fact, the properties of the tubes were more modest. Galileo's first tube magnified objects three times ($3\times$), while the one taken to Venice was $8\times$. Galileo decided, with his perfected tube, to push his request to the members of the *Signoria* (this may have been Sarpi's idea). On August 21st, the most respected people in Venice looked at the far quarters of the city from the *campanile* of St. Mark's Cathedral, and on August 24th Galileo triumphantly gave his tube to Leonardo Donato, Doge of Venice. Galileo did not skimp in publicizing his gift. He said that he extracted his idea "from the most secret considerations of perspective."

Many later said that Galileo overestimated his contribution or even appropriated a foreign invention for himself (Brecht's play talks about this). At least in his publications, Galileo always acknowledged that he constructed his tube having heard of the Dutch invention (but without detailed information and not having seen "the Flemish glass"). Later he stressed the originality of his approach: "Indeed, we know that the Fleming who was first to invent the telescope was a simple maker of ordinary spectacles who, casually handling lenses of various sorts, happened to look through two at once, one convex and the other concave,

---

[2] A translation of Galileo's accounts of his discovery appears in Stillman Drake, *Discoveries and Opinions of Galileo* (Garden City: Doubleday, 1957), copyright © 1957 by Stillman Drake. This quotation (p. 244) is taken from *Il Saggiatore* (*The Assayer*).—*Transl.*

[3] Ibid., pp. 21–52 passim, taken from *Sidereus Nuncius* (*The Starry Messenger*). Reprinted by permission of Doubleday & Co.—*Transl.*

and placed at different distances from the eye. In this way he observed the resulting effect and thus discovered the instrument. But I, incited by the news mentioned above, discovered the same thing by means of reasoning." The name "telescope" was proposed by Cesi (see below) in 1611, when Galileo demonstrated the tube in Rome; earlier Galileo had used the term *occhiale*.[4] One can assume that Galileo demonstrated the superiority of theory over practice: For many years no one was able to build tubes of comparable power (because of this, in particular, there was no confirmation of Galileo's astronomical observations).

Galileo's tube fulfilled its purpose: He was given an annual salary for life of a thousand florins, unheard of for a mathematician. Galileo was supposed to make twelve tubes for the *Signoria*, and to give none to anyone else.

The Starry Messenger

Soon Galileo had a 20× tube and wrote, "forsaking terrestrial observations, I turned to celestial ones." At the end of 1609 Galileo looked through his tube at the moon and discovered "that the surface of the moon is not smooth, uniform, and precisely spherical as a great number of philosophers believe it (and the other heavenly bodies) to be, but is uneven, rough, and full of cavities and prominences, being not unlike the face of the earth...." Moreover, Galileo turned his attention to the ashen light on the part of the moon not lit by the sun. He assumed this light to be "the reflection of the Earth." It turned out later that at the same time an Englishman, Thomas Harriot, and his student William Lower had begun to observe heavenly bodies by telescope (their observations were unknown to their contemporaries). Lower wrote in a letter to his teacher that the moon reminded him of a tart with jam that his cook had baked the previous week. Leonardo da Vinci and Mästlin, Kepler's teacher, had already spoken of the moon's ashen light.

Then, before Galileo's eyes, the Milky Way broke into separate stars: "All the disputes which have vexed philosophers through so many ages have been resolved...the galaxy is, in fact, nothing but a congeries of innumerable stars grouped together in clusters."

Finally, on January 7, 1610, Galileo turned his telescope towards Jupiter. Near Jupiter he discovered three stars. He did not doubt that he was seeing ordinary "fixed" stars, but something greatly attracted his attention. The following night, "unknowingly led by some sort of fate," he again looked at Jupiter. He had no reason to regret it! Again he saw the familiar stars, but...their position with respect to Jupiter had changed: Yesterday they were found on different sides of Jupiter but today all were on the same side. He could still assume the stars were fixed and explain

---

[4] spyglass, or eyeglass—*Transl.*

the change in relative location by Jupiter's motion. On January 9th "the sky was then covered by clouds everywhere." On January 10th and 11th he found only two of the three stars, but on the 13th, to the contrary, a fourth appeared.

Galileo saw a new solution: The stars he was observing moved in relation to Jupiter, they were its satellites—moons—and they disappear because they are eclipsed by Jupiter. By the end of the month he was sure, "passing from the sensation of enigma to the feeling of rapture." He wrote to Belisario Vinta, the Florentine Secretary of State: "But the greatest miracle of all is that I discovered four new planets and observed their own distinctive motions, and the differences in their motions relative to one another, and relative to the motions of the other stars. These new planets move around another big star in the same way Venus and Mercury and, possibly, other known planets move around the Sun." There is no doubt in what context Galileo viewed his discovery, but see how careful a formulation he still used in regard to "other known planets."

Until March 2nd, Galileo observed Jupiter's moons, taking advantage of each cloudless night, and as early as March 12th his famous *Sidereus Nuncius* (*The Starry Messenger*) appeared: "THE STARRY MESSENGER, Revealing great, unusual, and remarkable spectacles, opening these to the consideration of every man, and especially of philosophers and astronomers; AS OBSERVED BY GALILEO GALILEI, Gentleman of Florence, Professor of Mathematics in the University of Padua, WITH THE AID OF A SPYGLASS *lately invented by him*, In the surface of the Moon, in innumerable Fixed Stars, in Nebulae, and above all in FOUR PLANETS swiftly revolving about Jupiter at differing distances and periods, ..."

Everyday matters were superimposed on all this. It turned out that Galileo's salary would be increased only after a year and, moreover, his teaching duties began to weigh on him. He began to think about moving to Florence. The Grand Duke, Ferdinando I de' Medici had just died and Cosimo II, Galileo's former student, had ascended the throne. The duke's patronage could be unmatchable for solving many problems, especially in the difficult matter of defending the Copernican system. There was already no doubt that this would be Galileo's main work. He wrote in a letter to Vinta, in connection with the possible move: "The works which I must bring to conclusion are these. Two books on the system and constitution of the universe—an immense conception full of philosophy, astronomy, and geometry.[5]

Soon Galileo proposed, through Vinta, to name the moons of Jupiter the Cosimean or the Medicean stars, in honor of Cosimo de' Medici. The second form was chosen. The number of moons fortunately coincided with

---

[5] For a translation of this letter, see Drake, *Discoveries*, pp. 60–65.—*Transl.*

Cosimo's having three brothers. *Sidereus Nuncius* is dedicated to Cosimo de' Medici: "And so, most serene Cosimo, having discovered under your patronage these stars unknown to every astronomer before me, I have with good right decided to designate them by the august name of your family. And if I am first to have investigated them, who can justly blame me if I likewise name them, calling them the Medicean Stars, in the hope that this name will bring as much honor to them as the names of other heroes have bestowed on other stars?"[6] Later, all four satellites received their own names (Io, Europa, Ganymede, and Callisto), and in order to distinguish them from the moons of Jupiter that were discovered later, they will be called Galilean.

Galileo set off for Florence for the Easter vacation. He took a tube with him so that the duke could see "his" stars for himself. Galileo was surrounded by respect, a medal with the image of the Medicean stars was to be struck in his honor, the conditions for his move were roughly set, and only the title of Galileo's position remained to be specified. The ruler was pleased to have his name immortalized in the heavens; no other royal personage could boast of that. On May 14th, Galileo received a letter from France dated April 20th, in which he was asked "to discover as soon as possible some heavenly body to which His Majesty's name may be fitly attached," meaning Henri IV. It was specified that the star was to be called "rather by the name Henri than Bourbon." It turned out that the author of the letter rushed Galileo for nothing: As soon as he sent it, "the sovereign attended by fortune" was assassinated. Later Galileo wrote to Florence that the House of Medici was in an exclusive situation: Neither Mars nor Saturn turned out to have moons (in fifty years Huygens and Cassini discovered moons of Saturn, and later moons were also discovered around Mars).

Doubts plagued the Grand Duke. Rumors stubbornly spread that the stars given him were the fruit of Galileo's fantasies or the creation of his tube. Even Christophorus Clavius, the leading mathematician of the College of Rome, talked about it. The situation was made more complicated because no astronomer other than Galileo himself had seen the Medicean stars. Galileo paid for the fact that no one else had made such perfect tubes. Such an important discovery had to be verified by the three most famous astronomers: Kepler, Giovanni Magini, and Clavius. And for the time being the question of moving to Florence was put aside.

Kepler, Magini, and Clavius

Things were simplest with Magini. On the way from Florence to Padua, Galileo stopped at Bologna and showed Magini the stars he had dis-

---

[6] Ibid., p. 25.

covered. Magini, equally famous for his computational abilities and his guile, gave the impression that he could see nothing around Jupiter but had certainly been forewarned. He did not argue and was prepared to attribute it to his poor vision, but this could not comfort Galileo.

Kepler responded immediately to the report of the discovery. As early as April 19th, he wrote an enthusiastic letter to Galileo.[7] It turned out that the news about the new planets had already come to Germany in mid-March. Kepler gently scolded Galileo about the lack of an answer to his *Astronomia Nova...de Motibus Stellae Martis* (*A New Astronomy... Commentaries on the Motions of Mars*), which contained his first two laws and which he had recently sent him: "Instead of reading a book by someone else, [my Galileo] has busied himself with a highly startling revelation...of four previously unknown planets, discovered by the use of the telescope with two lenses."

The first reports were unclear. Kepler was afraid that Galileo had discovered new (more than six) planets in the solar system. Kepler strongly held to the opinion that there were exactly six planets, and that the number six was not accidental but related to the five regular polyhedra. Kepler's fantasy led to still another possibility: All the planets similar to the earth have one moon in common, and this is what Galileo must have discovered. "The earth, which is one of the planets (according to Copernicus), has its own moon revolving around it as a special case. In the same way, Galileo has quite possibly seen four other very tiny moons running in very narrow orbits around the small bodies of Saturn, Jupiter, Mars, and Venus. But Mercury, the last of the planets around the sun, is so deeply immersed in the sun's rays that Galileo has not yet been able to discern anything similar there." Kepler sought numerical laws everywhere! Then he thought about the fact that one can speak of planets revolving around "fixed stars," rather than the sun. He recalled Bruno's innumerable worlds and even thought about the possibility "that countless others will be hereafter discovered in the same region, now that this start has been made."

At the same time, Emperor Rudolf II received *Sidereus Nuncius* (Kepler was the Imperial Astronomer). Kepler unhesitatingly believed Galileo's report: "I may perhaps seem rash in accepting your claims so readily with no support from my own experience. But why should I not believe a most learned mathematician, whose very style attests the soundness of his judgment? He has no intention of practicing deception in a bid for vulgar publicity, nor does he pretend to have seen what he has not seen. Because he loves the truth, he does not hesitate to oppose even the most familiar opinions. . . ."

---

[7] This letter was published as *Dissertatio cum Nuncio Sidereo* (*Conversation with the Starry Messenger*). The following quotations are taken from the English translation by Edward Rosen, *Kepler's Conversation with Galileo's Sidereal Messenger* (New York: Johnson Reprint Corp., 1965), pp. 9–39 passim.—Transl.

It seems that the regularity of the distribution of the number of planetary moons bothered Kepler: "I should rather wish that I now had a telescope at hand, with which I might anticipate you in discovering two satellites of Mars (as the relationship seems to me to require) and six or eight satellites of Saturn, with one each perhaps for Venus and Mercury." Kepler did not know if he had an arithmetic or a geometric progression!

Kepler pointed out some of Galileo's predecessors (Mästlin spoke of the ashen light of the moon and Della Porta predicted the possibility of constructing an optical tube). Kepler hoped that the sun is brighter than the fixed stars, and wanted to believe in the uniqueness of our world: "This world of ours does not belong to an undifferentiated swarm of countless others." There is no limit to Kepler's fantasies: "It is not improbable, I must point out, that there are inhabitants not only on the moon but on Jupiter too.... Given ships or sails adapted to the breezes of heaven, there will be those who will not shrink from even that vast expanse."

Magini tried to draw Kepler to his side. Kepler was implacable: "We are both Copernicans—each is happy with his own kind." Critical remarks from Kepler's *Dissertatio* reassured Magini: "Now there remain only these four new servants of Jupiter to banish and destroy." Martin Horky, an astronomer in Magini's circle, initiated a series of pamphlets against Galileo in May 1610. In *Excursion Against the Starry Messenger*, he explained the moons of Jupiter as an optical illusion. Kepler's relations with Horky were equivocal. In a letter to Galileo he called the essay cheeky, and said he was "surprised at the impudence of this youth." To Horky himself, expressing surprise at his continuing doubts about "Galileo's stars," Kepler wrote: "...not surprised and do not accuse you; the opinions of those who philosophize must be free."

The absence of confirmation began to bother Kepler. He himself had no suitable tube. From Bologna came the university's conclusion (at Magini's instigation) that the stars could not be seen through Galileo's own tube. In August, the worried Kepler wrote to Galileo: "I cannot conceal from you that letters are arriving in Prague from many Italians that these planets cannot be seen with your optical tube...therefore I beg of you, Galileo, to send some witnesses as soon as possible....The entire proof of the truth of the observations lies on you alone." Happily, the emperor, Rudolf II, known not only for his caprice but also for his love of science, became passionate about the tubes. Finally, a sufficiently well-perfected tube appeared in Prague, and in September, Kepler saw the moons of Jupiter. The participants in the observation independently drew the positions of the stars and the drawings coincided. "Galileo, you have won!" exclaimed Kepler.

In September, Santini saw Jupiter's moons from Venice, and especially joyful news came in December: Clavius had seen the moons. True, he still was not "sure whether or not they are planets." In September, Galileo moved to Florence. He began a correspondence with Clavius (in the

Republic of Venice, it had been forbidden to correspond with Jesuits). "In truth, you, your Excellency, deserve great praise since you are the first who has observed them," Clavius wrote to Galileo. Galileo found the way to Magini's heart. He recommended his work on burning lenses to the Grand Duke and enabled him to obtain the vacated chair at Padua (Magini had sought this position even when Galileo moved to Padua from Pisa.) The careful Magini gave a positive opinion on Santini's testimony. One couldn't ask for anything more!

A Year of Great Discoveries

The year 1610, beginning with the discovery of the moons of Jupiter, was an unusually happy one for Galileo as an astronomer: He made almost all his remarkable astronomical observations in that one year. On July 25th, he again observed "Jupiter in the morning in the east, together with its retinue." After this he discovered "still another most unusual miracle." He communicated his discovery to Florence, asking that it be kept secret until he could publish it: "The planet Saturn is not merely one single star, but three stars very close together, so much so that they are all but in contact one with another. They are quite immovable with regard to each other. . . the middle star of the three is by far greater than the two on either side." Galileo sent a phrase to Kepler, encoded in the form of an anagram: "I have observed the most distant of the planets to have a triple form." Later, Galileo wrote: "So you see a guard of satellites has been found for Jupiter, and for the decrepit little old man [Saturn] two servants to help his steps and never leave his side."

Galileo did not reveal his secret for five months. Kepler and Rudolf had no patience to figure out the clue and made the most improbable conjectures: "As quickly as possible, satisfy our passionate desire to know what your new discovery consists of. There is no man whom you could fear as a rival." Galileo revealed his secret, adding that in the weakest tube Saturn looked like an olive. This is how Galileo's discovery first appeared in print, in the preface to Kepler's *Dioptrice* (*Dioptics*[8]), with the obligatory references.

Within two years, Saturn unexpectedly stopped appearing as a triple. Galileo attributed this to Saturn's motion around the sun and predicted that it could soon be seen again in the form of three stars. The prediction came true, but Galileo did not guess Saturn's secret. The secret was revealed in 1655, when Huygens, looking at Saturn through a $92\times$ telescope, discovered that Saturn is surrounded by a ring that at lower

---

[8] the branch of geometrical optics dealing with lenses and images. Kepler's remarks are also contained in the translation of *Sidereus Nuncius* by E.S. Carlos, *The Sidereal Messenger of Galileo Galilei* (London: Rivingtons, 1880), pp. 90–91. —*Transl.*

magnification appears as adjacent stars. The ring becomes invisible when the observer happens to be in its plane, and Galileo was lucky to see this rare phenomenon. The visual impression of Saturn evolved as telescopes became stronger, from an olive to a sphere surrounded by a ring. Huygens also discovered Saturn's largest moon—Titan.

Soon after Galileo sent Kepler his letter with the anagrammatic clue, there was news of still other planets. For a long time, Galileo had been intently observing Venus, both as a morning and as an evening star. There was a great deal of arguing over Venus and Mercury between the adherents of Ptolemy and Copernicus. The former could not agree where their "spheres" were—outside the sun's "sphere" or inside. For Copernicus' adherents it was clear that if these planets are dark bodies then, since they lie between the sun and the earth, they must at times be seen as partial discs (like the phases of the moon). This problem does not arise if we assume that the planets shine by their own light (Kepler's position, apparently) or that they are transparent (this possibility was seriously discussed). Perhaps the telescope could help see what had not been seen with the naked eye.

Castelli recalled this problem in a letter to Galileo on December 5, 1610: "Since (so I believe) Copernicus' position is correct that Venus revolves around the Sun, we clearly had to observe it sometimes with horns and sometimes not. . . , if, however, the small size of the horns and the emission of light do not prevent us from observing these differences." But Galileo hardly needed to be reminded of this. As early as December 10th, he sent a coded message to Kepler in Prague, via the Tuscan ambassador Giuliano de' Medici, about his discovery of the phases of Venus, together with an accompanying letter: "I am sending you a coded message about yet another of my unusual observations, which leads to the solution of the most important disputes in astronomy and which contains the most important argument in support of the Pythagorean and Copernican system." Kepler, as always, had no patience to figure out the clue: "You will see that you're dealing with a German among Germans!"

But the first to whom Galileo revealed his secret was Clavius. Galileo had just heard from Clavius that the astronomers of the *Collegio Romano* had also observed both the moons of Jupiter and the elongated form of Saturn. The support of the *Collegio* played a special role in Galileo's plans, and he hurried to surprise Clavius with his new discovery. Galileo described his observations of Venus after "its evening appearance," and talked about how its circular form unexpectedly became twisted to one side and turned towards the sun, so that Venus no longer looked like a semicircle but "became noticeably horned." He predicted the form Venus would take when it would be seen as a morning star, and concluded, "This is how, my lord, it is explained how Venus (and undoubtedly Mercury does the same) moves around the Sun, which without any doubt is the center of the greatest revolutions of all the planets. Moreover, we are sure that these planets by themselves are dark and shine only when illuminated by the Sun, which, as I think, does not travel with the fixed stars according to certain of my observations...." Clavius could have no doubt about where Galileo was heading! This brought Galileo's year of great astronomical discoveries to a close.

Galileo did not halt his astronomical observations, but they were basically a continuation of what he saw in 1610. He continued to observe the sunspots that had appeared in the summer of 1610, and by 1613 had discovered that the sun rotates on its axis; we have already spoken about how he saw the disappearance of Saturn's "appendages." At the end of his life, before he was finally blind, Galileo was fortunate to discover the appearance of the libration of the moon (as a result of which it became possible to observe more than half the moon's surface). But he would never again be able to devote as much time to perfecting the telescope or to astronomical observations. And the mysteries of the universe would not again reveal themselves to him as they had in that great year! Galileo's achievements were so great that it would be at least half a century before comparable discoveries would be made in observational astronomy (by Huygens and Cassini). Now other problems began to disturb Galileo, and to solve these problems it was important for him to go to Rome.

Subjugation by Rome

Galileo arrived in Rome on March 29, 1611. He enjoyed the special attention of the Grand Duke of Tuscany, arriving in the Duke's sedan-chair and staying at the Medici palace in Rome. The four astronomers of the *Collegio Romano*, Clavius, Grienberger, van Maelcote, and Lembo, received him warmly. Galileo discovered that the Jesuit fathers systematically observed the Medici stars through optical tubes, trying to determine their periods. On April 21st, one of the leaders of the Holy Office, Cardinal Roberto Bellarmino, sent him an official query "about the new celestial observations of a leading mathematician" (his name was not mentioned) regarding the Milky Way, Saturn, the moon, and the moons of Jupiter. The answer came on April 24th, essentially confirming the observations. He noted both substantial and minor discrepancies, e.g., not mountains on the "lunar body" but rather that its density is not uniform, and the stars forming Saturn did not seem to him to be separate.

On April 14th, Galileo became the fifth member of the *Accadèmia del Lincei* ("lynx-eyed"[9]), founded eight years earlier by Federico Cesi, Marquis of Montebello. The academy was devoted to the free study of nature, with no limitations. Cesi later wrote to Galileo, "Those whom we accept will not be slaves of Aristotle or of any other philosopher, but will

---

[9] referring to Lynceus the Argonaut, noted for his keen sight.—*Transl.*

be persons of noble and free thought in the study of nature." His friendship with Cesi played an important role in Galileo's later life, and he now put *Galileo Galilei Linceo* on his works. A demonstration of Galileo's surprising tube took place atop the Janiculum hill (that was when Cesi proposed to call it a telescope).

Galileo was also honored by the *Collegio Romano*. Odo van Maelcote read a report entitled, "The Starry Messenger of the Roman College." He called Galileo "the most remarkable and fortunate astronomer currently alive" and praised his discoveries, but gently said that Galileo's explanations of the phenomena he discovered are not the only ones possible. Galileo was given to understand the limits within which he must stay. This wish was expressed very precisely by Paolo Gualdo: "...you should be satisfied with the glory you have acquired thanks to observations of the moon, the four planets, and similar things, and not take on the defense of ideas so contrary to human reason...." Gualdo's advice also foreshadowed the path that Galileo would later take: "...many things can be uttered by way of disputation which it is not wise to affirm as truths, in particular, if against them you have general opinion, absorbed, if one can say, with the creation of the universe." Cardinal Bellarmino apparently also indicated to Galileo, during an audience, the limits of what was permitted. Bellarmino gave a more definite warning to the Tuscan ambassador Niccolini: "Galileo should stay within the indicated bounds, or else his works will be given to the theological experts for consideration," and the ambassador was to understand that nothing good could come of that.

The rest of Galileo's trip was successful. Cardinal del Monte wrote to the Grand Duke, "Galileo, during the days he was in Rome, gave much satisfaction and, I think, obtained much, for he was able to demonstrate his discoveries so well that all the worthy and leading people of this city recognized them not only as true and real, but also as staggering. If we now lived in the ancient Roman republic, I am convinced that they would erect a statue to him on the Capitoline, in order to pay respect to his exclusive prowess."

"Philosopher and First Mathematician to the Grand Duke"

Thus not even a year had passed before Galileo's surprising astronomical discoveries obtained recognition. Do not think, however, that the *Collegio Romano* stopped the accusations against him. As before, there were people who opposed the existence of new planets. There was still suspicion of optical tubes. The argument was most absurd (perhaps from today's standpoint). For example, Francesco Sizi reasoned this way: An optical tube is similar to eyeglasses, eyeglasses cannot fit young and old people equally well, both see the planets through Galileo's tube at the same time, and so it is an optical illusion. Also, in Pisa, Giulio Libri simply refused to look through the tube. "I hope that when he goes to Heaven he will finally see my moons, which he did not want to see from Earth," Galileo said after Libri's death. Many of Galileo's opponents understood that claiming his statements contradicted holy writ were especially effective denunciations to the Inquisition.

But if this is how things stood with phenomena that could be observed directly, then what dangers threatened Galileo for his statements supporting the Copernican system! In *Sidereus Nuncius*, Galileo promised to write *Massimi Sistemi*, in which "by six proofs and arguments of natural philosophy" he confirms that "the Earth moves and surpasses the Moon in its light."[10] His reconnaissance in Rome showed clearly that at the present moment these arguments found no support among "the leading lights." Galileo did not abandon his plans, but a long siege began. He understood very well that accepting Copernicus was not an internal scientific question, that he first had to convince the best scientists in the world, and that this required all his effort and would take him away from his immediate scientific activities. Many scientists doubted the validity of Galileo's solution. Einstein's opinion on this is well known: "As regards Galileo, I imagine him differently. We should not doubt that he passionately sought after truth, more than anyone else. But it is hard to believe that a man of vision sees sense in joining the truth he has found to the ideas of the superficial masses, confused in their petty interests. Was such a problem important enough for him to give it the last years of his life....Without needing to personally, he went to Rome to tilt there at priests and politicians. This picture does not correspond to my image of the aged Galileo's internal independence. I cannot imagine that I, for example, would have undertaken anything like that to defend the theory of relativity. I would have thought the truth was far stronger than myself, and it would have seemed to me ridiculously quixotic to defend it by the sword, astride Rosinante...." Galileo held to a different opinion, but he hardly seems a scientific Don Quixote. He did not tilt at "priests and politicians" as much as draw them to his side with the greatest art.

It is interesting to compare Einstein's statement with the view of the Pythagoreans, who were the first to accept a moving earth and fixed sun: "Let us only try to know something for ourselves, finding satisfaction just in this, and forget the desire and hope to rise in the eyes of the crowd or to strive for the approval of the bookish philosophers."

First of all, mathematicians traditionally were not supposed to discuss questions about the creation of the universe. To observe heavenly bodies, construct tables, use tables for horoscopes—these were the limits of a mathematician's duties. Galileo had no taste for making horoscopes (unlike Kepler, for instance) but at times he still had to do it. Thus, in

---

[10] i.e., in the amount of sunlight it reflects—*Transl.*

anticipation of his move to Florence and at the insistence of the Duchess, he had made a horoscope for Grand Duke Ferdinando I (the father of Cosimo II, the present Duke), who had fallen ill. The horoscope promised a favorable turn of events, the Duke was pleased, Galileo's son-in-law obtained a position he desired, and within several days the Duke died....You had to be at least a philosopher in order to discuss the creation of the universe (after all, even their salaries were noticeably higher than the mathematicians'), and if it conflicted with holy writ then you certainly had to be a theologian. Galileo could not become a theologian, but he could try to become a philosopher.

Galileo went through long negotiations over the title of his future position in Florence; he wanted the word "philosopher" in his title, for he had "studied more years in philosophy than months in pure mathematics." In the end, they agreed on the title "Philosopher and First Mathematician to His Highness the Grand Duke of Tuscany" (first mathematician, but not first philosopher!).

He began life in Florence in discussions with the conservative philosophers of the University of Pisa, followers of Aristotle, who assumed that the truth, "speaking in their own words, must be sought not in the universe and not in nature, but in the comparison of texts." Galileo was satisfied with his first successes: "How you, dear Kepler, would have laughed if you had heard how in Pisa, in the presence of the Grand Duke, the first philosopher of the local university came out against me, trying by arguments of logic, as if by bewitched incantations, to tear the new planets down from the heavens and destroy them!" His discussions concerned more than astronomy. In 1612, *Discorso Intorno alle Cose che Stanno in su l'Aqua* (*Discourse on Bodies Floating in Water*) appeared, devoted to hydrostatics and rather unpleasant for Aristotle's adherents. Within a year came *Istoria...intorno alle Macchie Solari...*(*Letters on the Solar Spots*), with barbs aimed in the same direction: "This news, I fear, will become the death knell or, rather, the death sentence for pseudophilosophy....I hope that the hilliness of the moon will become for the peripatetics a mere tickle compared to the torment of the clouds, steam and abundance of smoke that constantly arise, move about and disappear on the very face of the sun" (from a letter to Cesi; Aristotle's adherents were called peripatetics). Perhaps Galileo celebrated his victory prematurely....

Galileo was increasingly pulled into discussions with people who were far removed from actual science. Sometimes doubts plagued him: "With unspeakable disgust, I have reached this point and, as if I were repenting for my deeds, understood how fruitlessly I have squandered time and effort." The struggle intensified. The Dominican monk Tommaso Caccini, directing his sermons against Galileo, proposed radical measures: "Mathematicians must be banished from all Catholic states!" At the same time, Galileo decided to discuss theological questions. In 1614, copies were circulated of a letter he wrote to Castelli, in which one can find words such as: "Hence it appears that physical effects placed before our eyes by sensible experience, or concluded by necessary demonstrations, should not in any circumstances be called in doubt by passages in Scripture that verbally have a different semblance, since not everything in Scripture is linked to such severe obligations as is every physical effect."[11] This very letter probably served as the means for Father Nicolò Lorini's denunciation of Galileo to the Inquisition. It turned out that Galileo was accurate enough. The ravenous qualificators[12] could find only "three evil-sounding places" in the letter, and two of these were not in the original, which the Inquisition could not obtain.

In February 1615, a book appeared in Naples by Paolo Foscarini, a member of the Carmelite order, giving an account of the Copernican system in the form of a letter to the order's general. Bellarmino used the book as a way to state his relation to the problem, in a letter to Foscarini: "It seems to me that your Reverence and Signor Galileo act prudently when you content yourselves with speaking hypothetically and not absolutely, as I have always understood that Copernicus spoke. To say that on the supposition of the Earth's movement and the Sun's quiescence all the celestial appearances are explained better than by the theory of eccentrics and epicycles is to speak with excellent good sense and to run no risk whatever. Such a manner of speaking is enough for a mathematician. But to want to affirm that the Sun, in very truth, is at the center of the universe and only rotates on its axis without going from east to west, is a very dangerous attitude and one calculated not only to arouse all Scholastic philosophers and theologians but also to injure our holy faith by contradicting the Scriptures."[13] We must give the head of the Inquisition his due—he expressed his opinion with the utmost clarity.

In December 1615, Galileo was again in Rome. He probably wanted to influence the path of the investigation forming against him, and had not yet lost hope of changing the Church's opinion about the Copernican system.

A "Salutary Edict"

Galileo was in every way a diplomat. He visited Bellarmino, and tried to bring Cardinal Orsini over to his side. In a message to Orsini, he set forth his most secret argument in support of the earth's motion—the tides. He explained them by the mutual action of the daily and orbital motions of

---

[11] For an English translation of this letter, see Drake, *Galileo at Work*, pp. 224–229.—*Transl.*

[12] who examined cases and prepared them for trial—*Transl.*

[13] English translations, in whole or in part, of Cardinal Bellarmino's letter and other relevant documents appear in de Santillana, *The Crime of Galileo*, p. 99.—*Transl.*

the earth, and saw no competing explanation. Galileo thought up this explanation in Venice, where he saw how the water in a boat moved when it sped up and slowed down. "This phenomenon is indisputable, easy to understand, and can be verified by experiment at any time." Simpler explanations make the Copernican system very plausible, but a definite proof of the earth's motion can only be discovered on the earth itself! The future showed that Galileo's trump card was erroneous, but the explanation came much later. Galileo was at the very center of Roman intrigue: "I find myself in Rome, where just as the weather constantly changes, instability always rules in affairs."

Everything came to an end on February 24th, when a commission of eleven theologians voted that an assertion that the earth moved is "at least erroneous in faith." Galileo was told about this decision by the Commissary-General of the Inquisition in the presence of Cardinal Bellarmino. On March 5th, the Congregation of the Index "suspended" (but did not ban) Copernicus' book.[14] This act was almost symbolic. They planned to remove several phrases from the book about how the doctrine presented did not contradict writ, and to correct those places where Copernicus calls the earth a heavenly body (the sun and moon were heavenly bodies!) The Tuscan ambassador, in a letter to home, regretted Galileo's persistence but expressed the hope that he would not suffer. Rumors spread that Galileo would be required to swear an oath of renunciation, and Galileo obtained from Bellarmino reassurance refuting the rumors: ". . . but that only the declaration made by the Holy Father and published by the Sacred Congregation of the Index has been notified to him, wherein it is set forth that the doctrine attributed to Copernicus, that the Earth moves around the Sun and that the Sun is stationary in the center of the world, and does not move from east to west, is contrary to the Holy Scriptures and therefore cannot be defended or held."[15] This was in May before Galileo left Rome, and still earlier he had been received by Pope Paul V. What took place was not a sentence, but a stern warning. A violation of a clearly expressed ban was surely a crime.

Awaiting a Change

Galileo quit Rome, subject to the "salutary edict." But his obedience is not so apparent. Here, for example, is what he wrote while sending *Discorso sopra il Flusso e Reflusso del Mare* (*Discourse on the Tides*) to the Archduke Leopold of Austria, brother of the Grand Duchess: "Now, knowing as I do that it behooves us to obey the decisions of the authorities and to believe them, since they are guided by a higher insight than any to

[14] This was the "salutary edict" Galileo later referred to in his preface to *Massimi Sistemi.—Transl.*
[15] de Santillana, *The Crime of Galileo*, p. 132.

which my humble mind can of itself attain, I consider this treatise which I send you to be merely a poetical conceit, or a dream, and desire that your Highness may take it as such, inasmuch as it is based on the double motion of the Earth and, indeed, contains one of the arguments which I brought in confirmation of it."[16] It is hard to believe that this man will never say the earth moves. However, in order to return to this theme, Galileo needed not new arguments but a change in his everyday situation. And he waited for a change. Pope Paul V died, Giovanni Ciàmpoli, who was kindly inclined towards Galileo, became the influential secretary to the new Pope Gregory XV, and in 1621 the terrible Cardinal Bellarmino died. In 1623, Cardinal Maffeo Barberini, an educated man and patron of the sciences who did not hide his admiration for Galileo, became Pope Urban VIII.

At this time Galileo's pace quickened noticeably. In 1623, his book *Il Saggiatore* (*The Assayer*) appeared, devoted to comets and a response to Grassi (an astronomer of the *Collegio Romano*). Here he still did not speak directly about the earth's motion. But his following work, *Letter to Ingoli*, written in 1624, directly relates to this question. It was a reply to a 1616 essay by Francesco Ingoli, a highly educated clergyman, directed against the Copernican system. It is significant that Galileo waited eight years to respond. There are many brilliant and daring pages in this slim volume. There is even a poetic description of shipboard experiments that do not reveal the ship's motion, a remarkable explanation of the law of inertia; there are also arguments involving the fixed stars, comparing them to the sun and even a free discussion of the question of the size of the universe.

As for the latter, there is not even a hint of a universe bounded by "eight heavens" of fixed stars. Galileo clearly explains that he sees no arguments that allow one to choose between the hypotheses of a finite or infinite universe, but completely admits that only a small part is accessible to us: "...I am not at all sickened by the thought that the world, whose boundaries are set by our external senses, may turn out to be as small in relation to the Universe as the world of worms is in relation to our world." Galileo very daringly admits the hypothesis that the universe is infinite! Recall how uncomfortable the great fantasizer Kepler felt in assuming an infinite number of worlds similar to the solar system in his *Dissertatio*: "If you had discovered any planets revolving around one of the fixed stars, there would now be waiting for me chains and a prison amid Bruno's innumerabilities, I should rather say, exile to his infinite space."[17]

Letter to Ingoli was written in the autumn of 1624, and in the spring of 1625 Galileo again visited Rome. It seems that his goal was to make contact with the new pope, and to judge how favorable the situation had become. Galileo met with the pope six times, was treated very well by

---

[16] Ibid., p. 151.
[17] *Dissertatio*, p. 36.

Barberini's large family, and established favorable connections with many cardinals, including the influential German Cardinal Zollern. Relations with Galileo personally could not have been better, but his main hope was not justified: Urban VIII strongly supported the assertion of the "salutary edict" that the sun moves and the earth is stationary. Galileo discovered that in discussing this question, he and the pope had been speaking in different languages. Galileo claimed that the tides cannot be explained without assuming that the earth moves, but was told that what is unknown to people may be known to God. It is hard to argue with such reasoning! Galileo returned, and afterwards the pope sent a message to Grand Duke Ferdinando II (Cosimo had just died) expressing satisfaction with the Florentine scientist's visit and giving the most laudatory opinion of him.

Massimi Sistemi (World Systems)

Returning from Rome, Galileo finally decided to write a book setting forth all the arguments supporting the Copernican system. He had dreamt of this book in 1597 when he wrote to Kepler, had promised it in Sidereus Nuncius, and had considered it his main goal in moving to Florence. Galileo had turned sixty, and his health left something to be desired. The journey to Rome had not been a complete success, but there was no point in waiting for a better time. It would seem that after the "salutary edict" which, as we have explained, was strongly supported by Rome's "leading lights," he could not think of openly supporting the heliocentric system, but Galileo was not accustomed to guile.

Even in theological disputes, one of the participants was permitted to defend a heretical point of view "conditionally," so as to unmask it more graphically. The Copernican system was not declared heretical, and even Bellarmino had allowed it to be spoken of "hypothetically," as a mathematical construction. Galileo devised an artifice. Three interlocutors, Salviati, Sagredo, and Simplicio,[18] meet at Sagredo's palace and "dispassionately" discuss both world systems over the course of six days. The first two heroes are named for Galileo's deceased friends, and the third— an adherent of Aristotle and Ptolemy—is imaginary.

For more than five years, Galileo anxiously worked on the book; it goes without saying that he thought of it as the major work of his life. By 1630, four of the six days were finished: On the first day they discussed the possibility of the earth moving, on the second its daily motion, on the third its yearly motion, and, finally, on the fourth day the tides, Galileo's most beloved find. He decided to limit himself to four days, and to call the book Dialogo del Flusso e Reflusso (Dialogue on the Tides). In the spring of 1630, Galileo sent the manuscript to Rome.

---

[18] who later appear in Discorsi—Transl.

Galileo's book, in today's terminology, should really be called popular science. He consciously addressed it to the public at large, not just to scientists; he wanted to convince everyone that there were irrefutable arguments in favor of Copernicus. Partly because of this and partly because of his own scientific tastes, Galileo dealt almost exclusively with qualitative phenomena, without linking the system to the numerical data of astronomical observations. His planets moved uniformly in circles around the sun, which had no chance of agreeing with the observational data. In this regard, Galileo was significantly inferior to Kepler and avoided discussing the problems that bothered Copernicus. Evidently, computational astronomy was not Galileo's strong point.

Galileo obtained an audience with the pope, and met with the influential cardinals. Urban VIII was not against a book that would contain conditional arguments in support of a condemned system, but it could not create the feeling that the reader had a choice between two systems. The book must indicate unambiguously the finality of the assertion, sanctified by the Church, that the sun moves and the earth does not. Moreover, the pope rejected the title *Dialogo del Flusso e Reflusso*. Galileo promised to satisfy the pope's wish for a yet-unwritten introduction and conclusion. The manuscript was given to Niccolò Riccardi, the Master of the Holy Apostolic Palace (the chief licenser), also known as Padre Mostro, for an opinion. Padre Mostro chose a delaying tactic; unlike Galileo he was in no hurry.

The rest sounds like a detective story, with Galileo and his supporters acting with wonderful ingenuity so that the book would see the light of day. Ciàmpoli, the former papal secretary, evidently resorted to fraud just to obtain preliminary consent, risking his career. The book was supposed to be printed in Rome. With enormous cunning, with references to Galileo's health, plague in Italy, etc., it was printed in Florence.

On February 22, 1632, Grand Duke Ferdinando received a present, the first copy of the book dedicated to him: "A Dialogue of Galileo Galilei Linceo, Extraordinary Mathematician of the University of Pisa and Philosopher and Chief Mathematician of His Highness the Grand Duke of Tuscany, where in four days of meeting the two Grand Systems of the World of Ptolemy and Copernicus are discussed, and the philosophical and physical reasons for one side and the other are indefinitely propounded." The preface, addressed to the "discerning reader," explains the author's motives in presenting arguments supporting the Copernican system. He recalls the "salutary edict which, in order to obviate the dangerous tendencies of our present age, imposed a seasonable silence upon the Pythagorean opinion that the earth moves."[19]

---

[19] The quotations here and below are taken from the *Massimi Sistemi* section of Galileo Galilei, *Dialogue Concerning the Two Chief World Systems—Ptolemaic and Copernican*, trans. Stillman Drake (Berkeley: Univ. of California Press, 2nd rev. ed., 1967), pp. 5–6.—*Transl.*

Galileo's "zeal could not be contained" by the spreading rumors "that this decree had its origin not in judicious inquiry, but in passion none too well informed." The book must refute these rumors. He wants to show "foreign nations that as much is understood of this matter in Italy, and particularly in Rome, as transalpine diligence can ever have imagined. Collecting all the reflections that properly concern the Copernican system, I shall make it known that everything was brought before the attention of the Roman censorship, and that there proceed from this clime not only dogmas for the welfare of the soul, but ingenious discoveries for the delight of the mind as well." Finally, "it is not from failing to take count of what others have thought that we have yielded to asserting that the earth is motionless, and holding the contrary to be a mere mathematical caprice, but... for those reasons that are supplied by piety, religion, the knowledge of Divine Omnipotence, and a consciousness of the limitations of the human mind." In fact, his goals had to have been seen in Rome as worthy: to cut off talk about the rashness of the edict and to put the "foreign nations" in their place. Nevertheless, certain statements seem ambiguous today and may have also seemed so to some of the "leading lights." At the least, soon after copies of *Massimi Sistemi* appeared in Rome, there was a thunderclap.

Trial and Renunciation

The initiative for pursuing Galileo evidently came from Urban VIII himself. What angered the pope so and made him unappeasable? Perhaps he found the praise of the "seasonable salutary edict" insincere? *Massimi Sistemi* undoubtedly appeared at a very difficult time for Urban. The strong Spanish opposition in Rome was trying to remove the pope, and he could have been greatly threatened by accusations of supporting a scientist "suspected of heresy." It was said that the pope saw himself in the simpleton Simplicio, who defended the immobility of the earth. Galileo writes in the preface that this hero, unlike the two others, is not called by his proper name. What must Urban have thought if he really discovered something he had once told Galileo in Simplicio's verbiage?

In August 1632, the Papal Curia forbade the distribution of *Massimi Sistemi*. In September the matter was given to the Inquisition. A protracted game began. Galileo's supporters, including the Grand Duke, tried first to avoid consideration of the issue by the Inquisition, then to move the inquiry to Florence, and finally to procrastinate as much as possible, referring to Galileo's illness. All these attempts led nowhere— Urban VIII was implacable.

A threat to put Galileo in chains made him leave for Rome in January 1633. He arrived on February 13th, and on April 12th stood before Vincenzo Macolani, the Commissary–General of the Inquisition. An agonizing inquiry began, pressure was applied, and Galileo was apparently shown instruments of torture. An exhausting struggle to find a compromise took place. Three experts of the Holy See concluded that the book at least violated the ban on holding and spreading condemned doctrines. Galileo admitted that, against his wishes, he had strengthened the arguments in favor of the Copernican system. On June 22nd, in the monastery of Santa Maria sopra Minerva, the kneeling Galileo, who would reach seventy in half a year, heard the verdict. Because he "believed...that an opinion may be held and defended as probable after it has been declared and defined to be contrary to the Holy Scripture,"[20] Galileo was declared to be "vehemently suspected of heresy," and *Massimi Sistemi* was banned. Galileo was sentenced to "the formal prison of this Holy Office" (one suspected of heresy was not burned as a heretic!), and he must "for three years...repeat once a week the seven penitential Psalms." Then Galileo read out the text of renunciation: "...after it had been notified to me that the said doctrine was contrary to Holy Scripture—I wrote and printed a book in which I discuss this new doctrine already condemned and adduce arguments of great cogency in its favor without presenting any solution of these...." He swore to "fulfill and observe in all their integrity all penances" placed on him.

Perhaps at that moment Galileo regretted that he had abandoned the Republic of Venice, where he had been beyond the Inquisition's reach, and reassessed the Grand Duke's capabilities. But in Venice there had evidently been no chance of publishing his major work, which, regardless of the terrible consequences, he had succeeded in doing in Florence.

The Inquisition's prison sentence was replaced by exile, at first in the Medici palace in Rome; in two weeks, they sent him to Siena, to the archbishop Piccolomini. In half a year they decided to move him again, to his villa in Arcetri, near the convent where his daughters were. Galileo lived there for his eight remaining years, except for a short trip to Florence. Everywhere, he was under the vigilant eye of the Inquisition, which carefully controlled his contacts with the outside world. Urban VIII showed no pity to the disgraced scientist, even on the day of his death. His relative, Cardinal Barberini, wrote to Florence: "...it is not good to build a mausoleum for a corpse who was punished by the tribunal of the Holy Inquisition and died while under that punishment." The Grand Duke was unable to bury Galileo next to Michelangelo (this wish was fulfilled after many years).

Galileo's renunciation continues to disturb people even today. Did a scientist have the right to renounce a theory that he believed to be true without a doubt and to whose confirmation he had given a significant part of his life? Various explanations were put forth for Galileo's decision: The

---

[20] Complete English translations of Galileo's sentence and renunciation appear in de Santillana, *The Crime of Galileo*, pp. 306 ff.—*Transl.*

seventy-year-old ailing scientist's fear of torture and burning, the feeling that he had fulfilled his mission and that nothing could any longer interfere with the distribution of his book, and the possibility of preserving what proved to be his eight remaining years for scientific work (he returned to the studies he had interrupted for a quarter of a century, thanks to working out the ideas he now was forced to renounce). Constance Reid's book, *Hilbert*, tells what that great mathematician, with characteristic directness, said about Galileo: "But he was not an idiot. Only an idiot could believe that scientific truth needs martyrdom—that may be necessary in religion, but scientific results prove themselves in time." [21] Keep in mind that Galileo had also compromised before, and even after 1616[22] had formally acknowledged the immobility of the earth (and in *Massimi Sistemi* as well).

Galileo apparently never spoke the legendary phrase, *eppur si muove* ("still it moves"),[23] but regardless of his unquestionable faith his renunciation could not have been sincere. He must have been happy that *Massimi Sistemi* was not completely suppressed and that in 1635 a Latin translation appeared in Europe. Fulgenzio Micanzio, a Venetian acquaintance, wrote to him: "A remarkable thing—after your *Massimi Sistemi* came to light, people knowledgable about mathematics immediately went over to the side of the Copernican system. This is what the bans have led to!" Galileo answered: "What you have written to me about *Massimi Sistemi* is most unpleasant for me, since it can cause great concern among the leading lights. After all, the permission to read *Massimi Sistemi* is so limited that His Holiness keeps it only for himself, so that in the end it may happen that they will forget about this book completely."

The disgrace of the trial and verdict was difficult for Galileo, but so was the ban on continuing his work on the problems of the universe. He had no doubt that he should refrain from such work, but what was left for him? He had every reason to regret this period: "Our times are unhappy, a firm resolve to eradicate every new thought, especially in the sciences, now rules, as if everything possible to know were already known!" He could comfort himself with the predictions of the like-minded Tommaso Campanella in his *Defense of Galileo*, written in a Naples prison in 1616: "The coming century will judge us, for the present always crucifies its benefactors, but they later rise on the third day or in the third century."

Several weeks after the verdict, Galileo remembered the treatise on mechanics that had been cut short, and writing this book became his main endeavor for the coming years, the goal of his life. He recalled his youthful discovery of the isochronic property of the pendulum and entrusted his son Vincenzio with making a pendulum clock. Galileo's blindness was

---

[21] Constance Reid, *Hilbert* (New York: Springer-Verlag, 1970), p. 92.
[22] the year of the "salutary edict"—*Transl.*
[23] For more on this story, see Drake, *Galileo at Work*, pp. 356–357.—*Transl.*

inexorable. By the end of his work on the book he had lost his vision in one eye, but still looked at the sky from time to time through his telescope and described the libration of the moon, until at the end of 1637 he was totally blind: "...this sky, this world, and this universe, that with my startling observations and clear proofs I have extended a hundred and a thousand times compared to how the sages of all past centuries had usually seen it, has now so shrunk and narrowed for me that it has become no larger a space than is occupied by my own body. Because it has so recently occurred I still cannot regard this unhappiness with patience and resignation, but the passage of time should accustom me to it." Nevertheless, in the last year he was given, he again looked at the Medicean stars, and his old friends drew him to the idea that took possession of him in his last days.

The Medicean Stars Revisited

This idea may have occurred to Galileo even earlier, at the end of 1635 when he gave the French commission created by Cardinal Richelieu an opinion on Morin's method for determining longitude by observing the moon's motion. The method turned out to be unsound, but note how high-ranking a person was interested in it. The point is that finding the longitude aboard ship was one of the most pressing problems of the seventeenth century, the century of seafaring. Today it is hard to believe that at that time, sailors completed long voyages without any sort of reliable method for measuring the coordinates of a ship on the open sea. This of course did not apply to latitude, which, at least by the sixteenth century could be reliably measured (for example, by the sun's height at noon). Scientists could propose nothing workable for longitude. This problem worried the maritime powers, especially for economic reasons. The author of a method for measuring longitude with acceptable accuracy (say to half a degree) could at various times receive 100,000 écu from Philip II of Spain: or 100,000 livres from Louis XIV; or 20,000 pounds from the English Parliament; or 100,000 florins from the States General of Holland. Less accuracy decreased the prize proportionally. These numbers quite expressively demonstrate interest in the problem.

As long ago as the second century B.C., Hipparchus had an idea for measuring longitude: It used the fact that the difference in longitude between two points on the earth's surface is proportional to the difference in local time at these points. Thus at points whose longitudes differ by, say, 15°, the difference in local time equals one hour $(360°/24 = 15°)$. Therefore the problem can be reduced to measuring local time aboard ship and the corresponding time at some fixed point, for example, at the port from which it sailed. It is practical to measure the local time at the point where the ship is now, but how can we know the local time at the port? For a long time, no one even thought of "keeping" it. An excellent example is the story of the twenty-fours "lost" when Magellan sailed around the globe! And there were no clocks that could have kept that time, especially with the rocking motion of the sea.

Another possibility was to use astronomical phenomena that could be observed aboard ship and for which the precise time it would be observed in port was known. But few phenomena were suitable! What could be used, aside from solar and lunar eclipses, which are very rare? Tables for the moon's motion were so imperfect that longitude could not be measured from daily lunar observations (an example of such attempts was Morin's method). Galileo described the situation with a characteristic sense of triumph: "In former times, heaven was generous on that count, but for current needs it is rather stingy, assisting us only with eclipses of the Moon: and not because that same heaven does not abound with frequent phenomena that are visible and more suitable for our needs, but it has been convenient for the ruler of the world to conceal them right up to the present...." The optimism we feel in these words is connected to the hopes Galileo placed on the Medicean stars he had discovered, Jupiter's satellites. Among their peculiarities, discovered at the time of the first observations in 1610, are partial eclipses. If the moon's orbit were not inclined towards the earth's, the moon would fall in the cone of the earth's shadow at each full moon. Jupiter's moons fall in the broad cone of its shadow at each revolution, and they revolve rather quickly (Io completes a full revolution in about 42.5 earth-hours). While observing the eclipses of Jupiter's moons, Galileo decided to develop his own method for measuring longitude aboard ship.

Galileo began negotiations, without waiting to work out the method definitively. First he thought of Spain (it was probably important that this was a traditional Catholic country) and of meeting the viceroy in Naples, but gradually switched to Holland, where his idea aroused great interest. In 1636 secret negotiations with the States General were in full swing, and in August it was decided to ask Galileo for the necessary materials. Galileo wrote a triumphant message to the States General, the "tamer and ruler of the ocean." The quote given above was taken from this message. Galileo considered it symbolic that the telescope, which plays a leading role in his method, was invented in Holland. He did not stint in his description of the preeminence Holland would obtain through his method: "I could name a collection of arts, but it suffices to limit myself to seafaring, which has been brought by your Dutchmen to such startling perfection, and if the only remaining thing—the determination of longitude in which, as we see, they have so far been unsuccessful—joins the list of the rest of their clever operations, thanks to their recent and greatest invention, then their glory would reach such extremes that no other nation could dream of surpassing it."

A competent committee was formed, including Admiral Laurens Reael, the astronomer and mathematician Martinus Hortensius, and later Con-

stantijn Huygens, member of the Council of State and father of the
great scientist Christiaan Huygens. It was not easy for the practical Dutch
to believe the proposed method was feasible. "Imagine to how many
people of high position and wealth we were forced to preach a hitherto
unknown truth, that was first taken to be unreasonable," lamented
Huygens. Even the most supportive members of the committee were not
sure that the project could be realized. In a letter to Galileo, Admiral
Reael feared that his method might prove to be too refined "for so coarse a
people as the Dutch sailors." Doubts can be felt even in Huygens' words:
"Our people will with difficulty consider themselves indebted for a grand
gift that is more beautiful than profitable." Even Hortensius had difficulty
in adapting to seeing Jupiter's moons. Good telescopes were not enough.
At the end of 1637 Galileo sent his own telescope, which he could no
longer use because of his blindness. Tables were necessary to observe the
moons, and they were not easy to construct (for a long time not even the
periods of revolution could be determined).

Computational astronomy was never Galileo's strong suit, and now
blindness robbed him even of the ability to make observations. Galileo
asked the Olivetan monk Vincenzio Renieri, an experienced computational astronomer, to find the ephemerides of Jupiter's moons, at least
for the coming year. The calculations were delayed, and Renieri did not
succeed in constructing the tables that were needed.

The States General instructed Hortensius to meet with Galileo, to firm
up some necessary details, and to present him with a gold chain, as a gift.
At that point, the Inquisition interfered with the negotiations. A complicated game began, and as a result Galileo either thought it wise not
to meet Hortensius and accept the gift, or was directly prohibited to do so
by the Inquisition. Discussions began over keeping priority for Italy.
Castelli, who for a long time had not been allowed to see Galileo, even
received permission to meet with his teacher and learn the details of
the method. Hortensius and Reael unexpectedly died; Galileo's strength
deteriorated. The Florentine inquisitor informed Rome that the scientist,
"completely blind, will lie in his grave before he [again] studies mathematical constructions." Galileo did not lose hope, but it became clear that
he would not live to see the realization of his idea. In fact, it was probably
impossible to carry out the project. A long time went by before the
problem of measuring longitude at sea was finally solved, but in a completely different way—using accurate maritime clocks.

One of Galileo's last statements shows that he never stopped thinking
about the major question of his life, and speaks to his "incorrigibility":
"And just as I deem inadequate the Copernican observations and conjectures, so I judge equally, and more, fallacious and erroneous those
of Ptolemy, Aristotle, and their followers, when [even] without going
beyond the bounds of human reasoning their inconclusiveness can be very

easily discovered."[24] He was not allowed to argue against there being arguments inaccessible to man that refute Copernicus, but arguments accessible to man are enough to refute Ptolemy.

Epilogue

We see much about these events of three and a half centuries ago differently from Galileo. This involves both the difference between the Ptolemaic and Copernican systems and the question of the earth's "true" motion.

It is difficult to construct a consistent system of the universe that does not at heart rest on celestial mechanics. Paradoxically, Galileo's theory of celestial mechanics, as opposed to his "terrestrial" mechanics, was rather naive and close to Aristotle's view. First, he assumed that celestial bodies move because of inertia, rather than constantly acting forces. He had no acceptable notion of forces acting from afar and, for example, the idea of the solar or lunar attraction of terrestrial objects was considered an

[24] Ibid., p. 417.

astrological anachronism. Second, according to Galileo, celestial bodies moving by inertia exhibit uniform rotary motion. This is a *prima facie* contradiction of the "terrestrial" principle of inertia!

The main question for Galileo was the true (absolute) motion of the earth, and its experimental proof. Since terrestrial phenomena must be used for the proof, the terrestrial and celestial principles of inertia inexorably collide. With the greatest insight, Galileo refuted Tycho Brahe's claim, repeated by Ingoli, that phenomena aboard a moving ship must reveal that motion. Galileo's refutation (essentially the first statement of the "terrestrial" law of inertia) was mostly based on experiment. Simultaneously, he claimed that there are phenomena (the tides) that do reveal the earth's motion. How the hypothetical motion of the earth differs from a ship's motion, which cannot be discovered internally, is not clarified.

We emphasize that these phenomena were supposed to have been consequences of the earth's own motion, occurring by inertia without the participation of forces acting from afar. Galileo saw no contradiction here. As we have already noted, Galileo's "decisive" argument turned out to be completely wrong.

Galileo's view of true (absolute) motion was incorrect. The author of the law of inertia was still far from understanding the relative nature of motion and the role of a frame of reference. Christiaan Huygens did much to clarify this aspect of motion. Newton (unlike Huygens) assumed revolution was absolute. The Ptolemaic and Copernican systems use different frames of reference: The earth is fixed in one, and the sun in the other. The development of mechanics has shown that an opportunely chosen frame of reference is needed to reveal the laws of motion. The chief merit of the Copernican system was that it made the revelation of Kepler's laws (which, by the way, Galileo did not accept) possible. In Copernicus' system, the most massive body serves as a fixed origin, and as a first approximation in considering an individual planet we can restrict ourselves to the planet's interaction with the sun (its interaction with the other planets is negligible). This is the two-body problem, and Kepler's laws immediately follow from Newton's law of universal gravitation, as Newton showed. In a frame of reference where the earth is stationary, it becomes more complex to describe motion and, in particular, Kepler's laws do not hold.

Galileo's astronomical observations opened a new era in astronomy, and the moons of Jupiter played a special role. More than half a century passed before their periods were calculated, which Galileo himself and the astronomers of the *Collegio Romano*, who were experienced in calculations, had tried to do. Calculating their distances from Jupiter was even harder, because of insufficiently developed measurement techniques. But in 1685, when Newton published *On the System of the World*, part of his *Principia Mathematica* (*The Mathematical Principles of Natural Philosophy*), he could already say that Jupiter's moons obeyed Kepler's third law $T^2 \sim R^3$, where T is the period of revolution and R is the distance from Jupiter, although the data needed to be made more accurate. This was in the section *Phaenomena*, listing the experimental facts on which Newton's "world system" relied.

Constructing a theory of motion for the moons of Jupiter, based on the law of universal gravitation, tested the ambitions of the founders of celestial mechanics for a long time. A sufficiently precise theory needed to account not only for Jupiter's attraction, but also for the sun's and for the mutual attraction of the moons. In 1774 this problem was the theme for a prize given by the French Académie des Sciences.

Laplace constructed a rather precise theory in 1789. For a long time, the Medicean stars remained a goal that not one of the great astronomers could pass up. They presented the scientists with new and ever-surprising facts. Thus, for example, Laplace established that the time it takes for the first moon to revolve plus twice the time for the third is three times that of the second. But undoubtedly the most remarkable page in the study of Jupiter's moons is a discovery by Olaf Römer, which we will describe in detail.

Appendix: Olaf Römer's Conjecture

Cassini's Observations

Gradually, the telescope became a recognized astronomical tool. The power of telescopes grew: Christiaan Huygens' telescope gave 92-fold magnification, and in 1670 a telescope appeared in Paris that magnified objects 150 times. It is characteristic that this telescope was no longer at the disposal of a single scientist: It was installed at a new type of scientific institution, an observatory. The Paris Observatory, under the patronage of Louis XIV, was directed by Jean-Dominique Cassini (1625–1712), an Italian astronomer. Astronomy owes much to Cassini. He discovered that Saturn has four moons besides the one (Titan) discovered by Huygens, and the ring Huygens found around Saturn turned out, under Cassini's more careful observations, to consist of two rings separated by a gap (which came to be called Cassini's division). Cassini proved that Jupiter and Saturn rotate on their axes. He also did great work in the area of astronomical computation, measuring the astronomical unit, the distance from the earth to the sun, with an accuracy unheard of at the time. It is interesting to compare Cassini's value of 146 million kilometers with the true value of 149.6 million, and the 8 million that had been previously assumed.

As we have already noted, calculating the periods of revolution of Jupiter's moons became one of the central problems of astronomy in the second half of the seventeenth century. These values can be obtained by straightforward calculations if we know the successive times of their eclipses accurately. Conversely, knowing the periods of the moons, we can predict the times of their eclipses. In 1672 Cassini very carefully recorded the eclipses of Io (one of Jupiter's moons). He was surprised to find that his values for Io's period differed from time to time, as if the eclipse were sometimes a bit late and sometimes a bit early. The greatest difference he obtained was 22 minutes (for a period of 42.5 hours), and could not be explained by the accuracy of the measurement. Evidently, Cassini was already able to use Huygens' pendulum clock, which had begun to be used for astronomical observations. The observed effect had no reasonable explanation.

In 1672, the year Cassini systematically observed Jupiter's moons, a young Danish scientist named Olaf Römer (1644–1710) appeared at the Paris Observatory. He was intrigued by a striking coincidence (that Cassini may have also noticed): The greatest delay in Io's eclipse occurred at those times when Jupiter was furthest from the earth. It was possible that he noticed this phenomenon accidentally, but he must have had foresight not to explain it as an accident! Although at the time of Louis XIV the earth was still at the center of the universe in the astronomical atlases, scientists were not prepared to explain a change in the revolution of Jupiter's moon by the earth's influence! Römer proposed a competing explanation that must have seemed no less fantastic. He suggested that there is a delay in seeing Io's eclipse because its light travels a greater distance when the distance between the earth and Jupiter is greater. In order to evaluate Römer's hypothesis, we must recall what his contemporaries thought about the speed of light.

Digression on the Speed of Light

The ancient scholars assumed that light travels instantaneously (the only exception may have been Empedocles). For many centuries, this opinion was reinforced by Aristotle's authority. In the East, Avicenna and Alhazen[25] assumed the speed of light is finite but very large. Of the later European scientists, Galileo was one of the first who was ready to assume the speed of light is finite. In *Discorsi*, the interlocutors Sagredo, Simplicio, and Salviati discuss the problem.[26] Sagredo raises the question, and Simplicio assumes the speed of light is infinite since we see the flash of an artillery shot "without lapse of time." For Sagredo, the fact that the sound arrives after a noticeable delay means only that sound travels significantly more slowly than light. In response, Salviati, representing Galileo's interests in this triumvirate, proposes an experiment with two observers supplied with lanterns, where each one uncovers his lantern when he sees

---

[25] an Egyptian physicist and mathematician, al-Hasan ibn al-Haitham (c.965–1039) —*Transl.*

[26] *Discorsi*, pp. 42–44.

the other's light. But this experiment, which the scientists of the Florentine *Accademia* in fact tried to carry out, does not have a real chance of convincingly showing that the speed of light is finite. (Einstein and Infeld remark that for this they would have had to determine an interval of time on the order of 1/100,000 of a second.[27] Kepler assumed that light travels instantaneously; Robert Hooke thought its speed is finite but impossibly large to measure. Descartes and Fermat assumed it is infinite, which greatly complicated their research in geometric optics. Descartes assumed on the one hand that light travels instantaneously, but on the other hand decomposed its "speed" into components. Fermat, in trying to avoid talking about the speed of light while stating his famous principle of least time, resorted to every possible subterfuge, talking about "the antipathy of light towards matter" and introducing a formal coefficient that for all practical purposes is a ratio of speeds of light. Thus, most of Römer's contemporaries were not prepared to acknowledge the finiteness of the speed of light, let alone make it responsible for phenomena that were perfectly tangible but occurred on the astronomical scale. By comparison, we note that the speed of sound was only recently measured.

Römer's Calculations

Römer's calculations were of the utmost simplicity. He began with the fact that twenty-two minutes, the maximum delay in the onset of Io's eclipse, is exactly the time it takes light to travel a distance equal to the difference between the greatest and least distances between the earth and Jupiter. This difference is twice the distance between the earth and the sun. Compared to this, we may neglect the distance from Jupiter to Io.

We see that Römer had still another reason to be grateful to Cassini: a rather precise value for the distance from the earth to the sun (146 million kilometers). According to Römer, light thus requires 1320 seconds (22 minutes) to travel 292 million kilometers, and so he obtained 221,200 kilometers per second for the speed of light. One error lay in his value for the astronomical unit (the true value is 149.6 million kilometers), but the main error was a very large mistake in the maximum delay (the true value is 16 minutes and 36 seconds). For the correct values he would have obtained 300,400 kilometers per second for the speed of light, which is very close to the true value (299,792.5 kilometers per second). It is striking that Römer succeeded in obtaining a value of the correct order of magnitude.

Römer carried out these calculations in September 1676. To convince scientists he was correct, he thought of a stunt worthy of the ancient Egyptian priests. He carried out his calculations and predicted that in November, Io's eclipse would be ten minutes late. Observations, in which

---

[27] Albert Einstein and Leopold Infeld, *The Evolution of Physics* (New York: Simon and Schuster, 1938), p. 95.

Cassini took part, proved that Römer had accurately predicted the time to within a second. But this agreement did not make too great an impression on those around him, at least not the Parisian *Académie*, among whom the Cartesians (adherents of Descartes) predominated. After all, their teacher had written about astronomers that "although their propositions are always wrong and unreliable, they draw quite correct conclusions, relying on the various observations they make." Even Cassini refused to support Römer! This sort of thing is not at all rare in the history of science. Römer did have his adherents, including the English astronomer Edmund Halley (1656–1742).

Römer's theory was finally accepted when, in 1728, James Bradley (1693–1762) studied a visible annual motion of the stars—aberration. It had a natural explanation as the result of adding the speed of light leaving the stars to the speed of the earth in its orbit. Bradley found that the speed of light was 10,000 times that of the earth, which agreed well with Römer's figure. The fact that two essentially different paths led to the same answer convinced many people. The first measurement of the speed of light as the result of a "terrestrial" experiment was made by Armande Fizeau in 1849.

In telling about Galileo's discoveries today, we should not forget that the space probes Voyager 1 and Voyager 2 have enabled us to learn about the surfaces of Jupiter's Galilean moons. The probe that will be launched in 1989 especially to study Jupiter will carry Galileo's name. Jonathan Eberhart wrote about what scientists saw in the pictures that have been transmitted to earth: The Galilean satellites are "no mere collection of rockballs. Callisto, farthest of the four from Jupiter, presented perhaps the oldest, most heavily cratered surface yet studied. Ganymede...a whole gamut of tectonic thrashings, twistings, turnings and slippings. Europa amazed onlookers...smoother than a billiard ball, yet crisscrossed with myriad linear features that may be cracks left by global wrenchings but which somehow survived through the eons in the icy crust. And finally, stunning Io, bedecked in red and gold, silver, black and white, seething with sulfurous volcanic activity that is one of the major discoveries in the history of planetary exploration. A whole, previously unimagined family of exotic worlds, each radically different not only from its companions, but also from everything else in the planet-watcher's experience."[28]

---

[28] *Science News*, April 19, 1980, p. 251. Reprinted with permission from *Science News*, the weekly newsmagazine of science, copyright 1980 by Science Service, Inc.

Christiaan Huygens, Pendulum Clocks, and a Curve "Not at All Considered by the Ancients"

We have told how Galileo laid the foundation for classical mechanics almost at the beginning of the seventeenth century. Christiaan Huygens (1629–1695) was Galileo's immediate scientific successor. In Lagrange's words, Huygens "was destined to improve and develop most of Galileo's important discoveries."[1] There is a story about how Huygens, at age seventeen, first came into contact with Galileo's ideas: He planned to prove that a projectile launched moves horizontally along a parabola, but discovered a proof in Galileo's book and did not want "to write the *Iliad* after Homer." It is striking how close Huygens and Galileo were in scientific spirit and interests.

It sometimes seems that a rejuvenated Galileo was again perfecting his optical tubes and continuing the astronomical observations he had interrupted forty years before. He tried to use the most powerful telescope to guess the secret of Saturn, which appears as a trio of joined stars, and finally, looking through a 92× telescope (Galileo's was 20×), discovered that the adjacent stars are Saturn's ring. He returned again to the problem that was of such keen interest in 1610: Do any planets besides the earth and Jupiter have moons? At that time Galileo wrote the Medicis that no moons had been discovered around the other planets, and that no royal house except for Medici, in whose honor he had named the moons of Jupiter, could claim its "own" stars. Huygens discovered Titan, Saturn's moon, in 1655. Probably, times had changed: Huygens did not offer the moon he discovered to anyone as a gift.

Then Huygens turned to mechanics, where he was concerned with the same problems as Galileo had been. He developed his principle of inertia, stating that not only is it sometimes impossible to discover motion by internal means, but that the very assertion that the body moves has no absolute meaning. Huygens understood every motion as relative, which was quite unlike Newton's view. Galileo, reflecting on why a body stays on

---

[1] *Mécanique Analytique*, p. 207—*Transl.*

Christiaan Huygens (contemporary engraving), 1629–1695.

the earth's surface during its rotation, once almost obtained the formula for centripetal acceleration, literally not making the last step (see page 39.) Huygens completed Galileo's argument and obtained one of the most remarkable formulas in mechanics.

Huygens then turned to studying the isochronous nature of the oscillations of a mathematical pendulum. This was probably Galileo's first discovery in mechanics, and here too Huygens was able to add to what Galileo had done: A mathematical pendulum turns out to be isochronous (the period of oscillation of a pendulum of fixed length is independent of the amplitude of its swing) only approximately, for small angles of swing. Finally, Huygens brought to fruition the idea that occupied Galileo in his last years: He constructed a pendulum clock.

Christiaan Huygens worked on the problem of creating and perfecting clocks, especially pendulum clocks, for nearly forty years, from 1656 to 1693. Arnold Sommerfeld called Huygens "the most brilliant expert on clocks of all time." One of Huygens' basic memoirs, containing his results in mathematics and mechanics, appeared in 1673 under the name *Horlogium Oscillatorium* (*Pendulum Clocks, or Geometric Proofs Relating to the Motion of Pendula Adapted to Clocks*).[2] Huygens invented much in trying to solve one of the most fundamental problems of his life—creating a clock that could be used as a marine chronometer; he thought through many things from the standpoint of their application to this problem (cycloid pendulums, the theory of developments of curves, centrifugal forces, etc.). We will talk about Huygens' research in chronometry, sticking to the mechanical and mathematical problems that concerned him. But first of all we should explain why a problem about clocks attracted a great scientist.

Clocks are among the most ancient human inventions. At first there were solar, water, and sand clocks; in the Middle Ages, mechanical clocks appeared. The measurement of time played different roles in people's lives in different eras. The German historian Oswald Spengler, noting that mechanical clocks were invented at the emergence of the Romanesque style and of the movement leading to the crusades, writes: "... the mechanical clock, the dread symbol of the flow of time, and the chimes of countless clock towers that echo day and night over West Europe are perhaps the most wonderful expression of which a historical world-feeling is capable. In the timeless countrysides and cities of the Classical world, we find nothing of the sort. . . . In Babylon and Egypt water-clocks and sun-dials were discovered in the very early stages, yet in Athens it was left to Plato to introduce a practically useful form of clepsydra. . ."[3] [a variety of water-clock—*S.G.*]

It is typical that in the first steps of the new mechanics and mathematical analysis, time did not immediately take the role of a fundamental variable quantity in the description of motion (in his search for the law of free fall, Galileo began with the hypothesis that velocity is proportional to distance, rather than time).

For a long time, mechanical clocks were inconvenient and imperfect. Several methods had been invented for transforming the accelerated fall of

---

[2] This appears, with a French translation, in Huygens' collected works, *Oeuvres Complètes*, ed. Société Hollandaise des Sciences (The Hague: Nijhoff, 1938), vol. 18.—*Transl.*

[3] *The Decline of the West*, trans. Charles F. Atkinson, copyright 1926 by Alfred A. Knopf, Inc., vol. 1, pp. 14–15.

a weight into the uniform motion of a pointer, but even Tycho Brahe's astronomical clock, whose accuracy was known, had to be "adjusted" with a hammer every day. Not a single mechanical phenomenon was known that would periodically repeat itself in a relatively small fixed amount of time.

Pendulum Clocks

Such a phenomenon was discovered at the dawn of Galileo's creation of the new mechanics. Namely, Galileo discovered that the oscillations of a pendulum are isochronous, i.e., the period, in particular, does not change as the oscillations are damped. We have earlier described Viviani's story of Galileo's discovery.

Galileo proposed to use a pendulum to make a clock. In his letter of June 5, 1636, to Admiral Reael, he wrote of joining a pendulum to an oscillation counter. However, he began work on making a clock in 1641, a year before his death, and did not finish. His work was supposed to have been continued by his son Vincenzio, who was slow in renewing it and began only in 1649. This was not long before Vincenzio's death, so he too did not succeed in making a clock. Various scientists had already used the isochronous property of the pendulum in laboratory experiments, but it was not an easy path from there to creating pendulum clocks.

It was accomplished in 1657 by the twenty-seven-year-old Christiaan Huygens, already a well-known scientist because of his discovery of Saturn's ring. On January 12, 1657, he wrote: "During these days I have found a new way of making clocks, with which time can be measured so precisely that there is no little hope of being able to use it to measure longitude, even if this is to be done at sea." The first example of a pendulum clock was made by the Hague watchmaker Salomon Coster, and on June 16th the States General of Holland issued a patent, strengthening Huygens' priority. In 1658, he published a description of his invention in *Horologium* (*The Clock*).

Learning of Huygens' clock, Galileo's students undertook an energetic attempt to establish their teacher's priority. In order to appreciate the situation, it is important to understand that in the seventeenth century the problem of making accurate clocks was associated, first of all, with the possibility of using them to measure longitude aboard ship. Galileo understood this, and Huygens set it forth from the very beginning (as the above quotation shows).

We have already discussed the problem of measuring longitude. Galileo's students knew that at the end of his life he had carried out secret negotiations with the States General, proposing his method for measuring longitude. The contents of the negotiations, interrupted after the Florentine Inquisitor's interference, were not reliably known. One could assume that there was talk of using pendulum clocks. Recall that the idea for this method consists of the fact that clocks "remember" the time at the port of departure, and the difference between this time and local time aboard ship determines the difference in longitude. It was important that a clock keep time correctly when being tossed about by the waves. A pendulum's isochronous oscillations would be essential because of both damped oscillations and rough seas.

Galileo suggested to Holland another way to measure longitude, based on observations of the eclipses of Jupiter's moons. Although pendulum clocks may have been mentioned in the negotiations (the letter to Reael), undoubtedly no construction for clocks or detailed information about them were given to Holland. When Galileo began working on making a clock (1641), the negotiations with the States General had practically ceased.

Huygens was not accused of plagiarism, although people may have been aware that pendulum clocks were being made in Holland, by the son of an influential member of the Council of State who had been involved in the negotiations with Galileo. Leopoldo de' Medici wrote a letter to the French astronomer I. Boulliau, who protected Huygens, with a commission to make a working mechanism following Galileo's idea. Viviani's story, mentioned earlier, and a drawing of Galileo's clock were attached to the letter, to be given to Huygens. Huygens, familiarizing himself with the drawing, said that the basic idea was there but not its technical realization. In 1673, Huygens wrote: "...if they say that [Galileo] tried to find this machine but without being able to reach his goal, it seems to me that they diminish his glory more than mine, since in this case I searched for the same thing as he with more success."[4] Here it is unnecessary to recall that Galileo worked with clocks when he was blind and more than fifty years older than Huygens was when he worked on the same problem.

Huygens' first clocks mostly employed the construction of those in use at the time (he had in mind being able quickly to remake the clocks he already had into pendulum clocks). From that time on, perfecting clocks became one of Huygens' chief concerns. His last work on clocks was published in 1693, two years before his death. If in his first work Huygens appeared most of all as an engineer, knowing how to realize the already-known isochronous property of the pendulum in a clock mechanism, then gradually Huygens the physicist and mathematician came to the foreground.

Incidentally, his engineering achievements were outstanding in number. The German physicist Max von Laue (1879–1960), who won the Nobel Prize for his 1912 discovery of X-ray crystallography, highlighted the idea of feedback in Huygens' clocks: The initial energy is communicated to the pendulum without altering its period of oscillation, "and then the very source of the oscillations determines the moments of time when additional energy is required." For Huygens this role was fulfilled by a simple and

---

[4] *Oeuvres*, vol. 18, p. 90.

clever construction in the form of an armature with teeth cut slantwise, rhythmically nudging the pendulum.

While still beginning his work, Huygens discovered an inaccuracy in Galileo's assertion that the oscillations are isochronous. A pendulum has this property only for small angles of deviation from the vertical, but not, say, for an angle of 60° (Galileo may have considered this in the experiments described by Viviani). In 1673 Huygens noted that the ratio of the period for 90° to that for small angles is 34/29.

A Tautochrone

In order to compensate for deviations from isochronicity, Huygens decided to decrease the length of the pendulum when the angle was increased. For this purpose, he employed restrictions in his first clocks, in the form of "cheeks" on which threads were wound, leading to pendants. Huygens did not establish an empirical method for choosing the form of the cheeks. In 1658 he removed them completely from the construction, introducing instead restrictions on the amplitude. But this did not mean he had stopped looking for an isochronous pendulum. Correcting discs appeared again in the clocks of 1659, but by this time Huygens was able to determine the form of the cheeks theoretically.

Here is how he solved the problem. Instead of a moving pendulum whose length decreases as it gets further from the vertical, consider a point mass moving along a trough in the form of the curve along which the end of the pendulum moves (for a mathematical pendulum this is a circle). Thus we have to find a curve (called an *isochrone* or *tautochrone*) for which the point reaches the bottom at the same time, regardless of the height from which it begins. Galileo erroneously assumed that a circle has this property (his claim that a mathematical pendulum is isochronous can be restated in this way). Huygens, though, discovered that a cycloid is a tautochrone, and by a happy circumstance the search for an isochronous pendulum coincided with serious research into cycloids for a different reason.

A fixed point on a circle that rolls along a line without slipping describes a cycloid. Galileo discovered the cycloid and gave it this name ("coming from circle"); it was called a trochoid or *roulette* in France (it was evidently discovered independently there by Marin Mersenne). Blaise Pascal wrote:

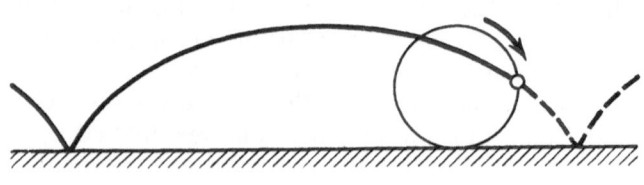

Cycloid.

"The *roulette* is so common a curve that after the line and circle no other is so often met; it is described so often before everyone's eyes that it is surprising it was not at all considered by the ancients...because it is nothing other than the path made through the air by the nail on a wheel, rolling as usual, from when the nail begins to leave the ground until the continuous rolling of the wheel has brought it back to the ground after a complete revolution."[5,6]

When the cycloid was discovered, it quickly became a very popular curve among mathematicians. In 1673, Huygens stated that the cycloid was studied more closely and intently than other curves. Mathematicians at the time were creating general methods for studying curves and were in great need of experimental material. Each new method had to be tested on the cycloid, which did not resemble the usual algebraic curves. For example, the cycloid is supposed to have solved a dispute between Pierre Fermat and René Descartes on the superiority of the methods they had proposed for extending tangents.

The kinetic definition of the cycloid lent itself to elegant solutions of various problems. Huygens' discovery was based on the properties of the tangent to a cycloid. Following Torricelli and Roberval, we can construct the tangent using the fact that the cycloid is the trajectory obtained by adding rectilinear motion along a direction line to the rotation of the rolling (generating) circle. The velocity vector of this motion is directed along the tangent, and is the sum of the velocity vectors of the constituent motions.

Thus, if A is the position at some moment of time of the point being observed, then we must add a horizontal vector to the vector tangent to the generating circle at A. Their lengths must be equal, because the circle rolls without slipping. This means we must construct a rhombus with vertex A, with one side horizontal and the other tangent to the circle, and form the diagonal of the rhombus (the lengths of the sides do not affect the direction of the diagonal). To this end, we construct a parallelogram $ABCD$, with sides AB, AC directed as shown, and with vertex D at the top of the circle. Then the right triangles ABO and BDO, where O is the center of the circle, are congruent, i.e., $AB = BD$, so we have a rhombus. Thus *at each point of a cycloid, the line joining that point to the point that is then at the top of the generating circle is tangent to the cycloid*. Note that the line joining A to the point E at the bottom of the circle is normal to the cycloid, i.e., perpendicular to the tangent. Huygens' attention was drawn to the cycloid thanks to an invitation to enter a competition to solve six problems concerning cycloids, announced by Pascal in June 1658. We will talk about

---

[5] *Histoire de la Roulette*, in *Oeuvres Complètes*, Jacques Chevalier, ed. (Paris: Gallimard, 1954), p. 194.
[6] Nicholas of Cusa considered the curve described by a "nail on a wheel" in the mid-fifteenth century.

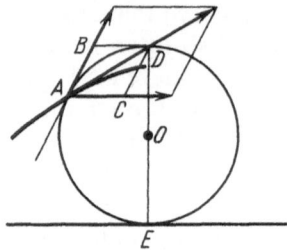

A tangent to the cycloid.

the competition in the chapter devoted to Pascal. In the short time available to the contestants Huygens solved four of the problems, more than anyone else except Pascal himself, who participated under the pseudonym Amos Dettonville.

After the competition Huygens returned to thinking about an isochronous pendulum. He studied how a point mass rolls along an "upside down" cycloid. Let r be the radius of the generating circle and let the point roll from a height of $H \leq 2r$. Let $h(t)$ be the height of the point at time t; $h(0) = H$. The magnitude of the velocity is determined from the law of conservation of energy and equals

$$|v(t)| = \sqrt{2g(H - h(t))};$$

the velocity is directed along the tangent to the cycloid. We find the vertical component of the velocity using the above rule for extending the tangent. If, at an arbitrary point A, we extend the tangent AD and let C be the

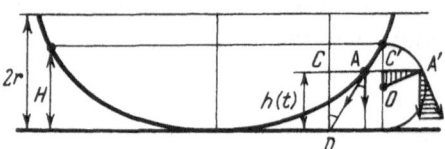

projection of A onto the vertical, then $|CD| = h(t)$. In this construction, we have

$$v_{\text{vert}} = |v(t)| \cos ADC \qquad h(t) = 2r \cos^2 ADC$$

$$\cos ADC = \sqrt{\frac{h(t)}{2r}}$$

$$v_{\text{vert}} = \sqrt{\frac{g}{r}} \cdot \sqrt{h(t)(H - h(t))}.$$

Now we can forget about the motion of a point on the cycloid and study the rectilinear motion $h(t)$ with velocity $v_{\text{vert}}(t)$ and initial condition $h(0) = H$. We must find the value $t = \tau$ for which $h(\tau) = 0$. This is a typical problem in differential equations, but Huygens thought of a trick. He considered yet another auxiliary motion: Let a point travel with velocity w, uniformly around a circle of diameter H (rather than $2r$), starting at the top. Suppose that at time t it is at a point A' with height $h(t)$. It is not hard to find the vertical component of the velocity at this point. Indeed,

$$w_{\text{vert}} = |w| \cos C'A'O$$

$$\cos C'A'O = \frac{|C'A'|}{OA'} = \frac{2|C'A'|}{H} = \frac{2}{H}\sqrt{h(t)(H - h(t))}$$

$$w_{\text{vert}} = \frac{2|w|}{H}\sqrt{h(t)(H - h(t))},$$

where O is the center and C' is the projection of A' onto the vertical diameter. If $2|w| = H\sqrt{g/r}$, then the vertical projection of the point revolving around the circle will move in the same way as the vertical projection of the point rolling along the cycloid. In particular, both points reach the bottom at time $\tau = \pi\sqrt{r/g}$. Here H cancels out, which reflects a remarkable fact: *the time τ for a point rolling along a cycloid to reach the bottom is independent of the height H at which it begins, and equals $\pi\sqrt{r/g}$. Thus, the cycloid is a tautochrone.*[7]

A point mass rolling along a cycloidal trough returns to its initial position in time $T = 4\tau$; T will correspond to the period of oscillations of a cycloid pendulum. We have

$$T = 4\pi\sqrt{\frac{r}{g}} \qquad (*)$$

Formula (*) recalls Galileo's hypothetical formula for the period of a mathematical pendulum of length l ($T = 2\pi\sqrt{l/g}$), so it was natural to try to use (*) to establish the latter. And indeed, Huygens used (*) to obtain the first rigorous proof of Galileo's formula for small swing angles φ. He noted that for small angles, a circular trough is almost the same as a cycloid, and it remained only to find the relation between the length l of a mathematical pendulum and the parameter r of a cycloid that minimizes this difference. This turned out to be $l = 4r$ (this is not obvious, and we will return to it later). Substituting $r = l/4$ into (*), we obtain the formula for the period of a mathematical pendulum: $T \approx 2\pi\sqrt{l/g}$ (for small φ).

The Cycloid Pendulum

This did not completely solve the problem of an isochronous pendulum. It showed the end of the pendulum had to move along a cycloid, but not how

---

[7] We have actually shown that the motion of a point mass in a cycloidal trough can be represented as the sum of a uniform rotary motion with angular velocity independent of the initial height H of the point, together with some (in general, nonuniform) translation motion. For $H = 2r$ it is easy to deduce this from the kinematic definition of the cycloid.

to make it happen. To this end cheeks were used, on which the string was wound; their form had to be found.

In *Horlogium Oscillatorium*, this problem was solved as part of the general problem of developing curves. It is noteworthy that Huygens became interested in these questions as early as 1654, long before his research on the isochronous pendulum.

Suppose we have an obstacle bounded by a curve L, and attach a stretched string of length l at some point O of L. We wind the string around the obstacle, keeping it stretched, and observe the curve M described by its free end. Huygens called the curve M the *development*[8] of L; now M is called the *involute* of L and L is called the *evolute* of M (with one evolute we associate many involutes, corresponding to different lengths l). We must find the evolute of the cycloid.

The curve M consists of those points B for which the sum of the lengths of the tangent BA to L at A and of the arc AO of L equals l (this corresponds precisely to stretching the partially wound string). Huygens first conjectured that *the tangent to M at B is perpendicular to AB*, i.e., that AB, the tangent to L at B, is simultaneously normal to M at B. The simplest way to explain this is from the kinematic definition of M. Recall that the velocity vector is tangent to the trajectory of motion and that as the action of the force changes, the velocity vector cannot change instantaneously. We "chop off" the obstacle at A, but continue the motion of the stretched string; then the end of the string begins to move along a circle with center A. Its velocity vector at B does not change, so at B the curve M and the circle with center A will have the same tangent, perpendicular to the radius BA.

Huygens next guessed that in a "good" situation, *the evolute of a curve can be uniquely reconstructed* (recall that one curve has many involutes)! The point is that the normals to M are the tangents to its evolute L and a "good" curve can be reconstructed from its tangents. Taking several tangents, we construct the polygonal line they describe; taking more and more tangents, we approximate the curve better and better (we say the curve is enveloped by its set of tangents).

We must find the curve whose tangents will be normal to a given cycloid. Huygens conjectured that *this curve will be the same cycloid, raised by 2r and shifted by half a period* (so that its troughs will fall on the cusps of the original cycloid).

Indeed, let $r = 1$, d and d' be the directrices of the lower and upper cycloids, respectively, and O and O' be their initial points (d' is two units above d and O' is π units to the right of O). Take a point C on d and consider the generating circles of the cycloids when they are each tangent to d at C. Let C' and C'' be the points diametrically opposite C on the

---

[8] Ibid., p. 188.

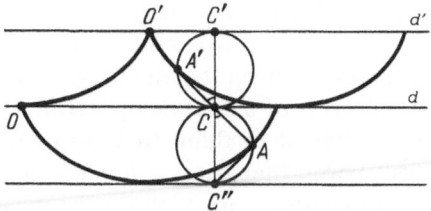

upper and lower circles, respectively, and A and A' be the corresponding points on the cycloids. The arc $CC''A$ is equal in length to the segment OC, so it is π units greater than the arc $C'A'$, whose length is that of $O'C'$. So the angles $C'CA'$ and $C''CA$ are equal, and the points A', C, and A are collinear. It remains to note that CA' is tangent to the upper cycloid and CA is normal to the lower one (AC'' is its tangent).

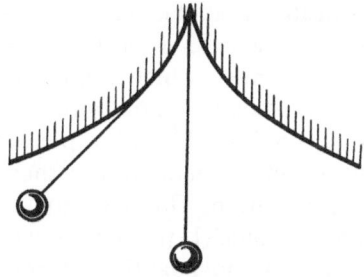

Now we know that the "cheeks" of a tautochronous pendulum must be cycloids and the length l of the string must equal $4r$ (for this value of l the involute is the cycloid we need). For small swing angles φ, the regulating "cheeks" have almost no influence on the length of the pendulum, and the cycloid is nearly a circular arc of radius $4r$ (see the end of the preceding section).

Christopher Wren's Theorem

The string is completely wound when its end is at the intersection of the two cycloids. Thus the length of one cycloid arc is twice the length of the string, i.e., $8r$. This theorem, which for Huygens was a simple corollary of the theory of developments of curves, was proved by the English mathematician[9] Christopher Wren in 1658, in connection with Pascal's competition.

Wren's theorem made a great impression on his contemporaries. Even after mathematicians had achieved great success in finding the areas of

---

[9] and architect, who rebuilt St. Paul's Cathedral in London—*Transl.*

curvilinear figures, they had made no progress on the rectification problem of constructing, with straightedge and compass, an interval whose
length is that of a given curve, or on the algebraic rectification problem of
expressing the length in terms of algebraic operations. By the mid-seven-
teenth century they had begun to think that, in general, rectification is
impossible (Descartes' words that "we, human beings, cannot find the
relation between lines and curves" are sometimes interpreted in this way).
Wren's rectification of the cycloid refuted this point of view. For a time it
was thought that what was involved was that the cycloid is not an algebraic
curve, but William Neile, Hendrik van Heuraet, and Pierre Fermat independently discovered that the semicubical parabola $y^2 = ax^3$ admits an
algebraic rectification (Neile's work even preceded Wren's, but was not
widely known).

Huygens' theory revealed the seemingly mysterious reason why the
semicubical parabola has this remarkable property. It turns out that its
involute is the usual quadratic parabola. More precisely, the evolute of the
parabola $y = x^2$ is the curve $y = \frac{1}{2} + 3(x/4)^{2/3}$. Huygens systematically
thought through the consequences, which yield a theory of developments
of curves beyond their application to pendulums: ". . . in order to apply this
property [that the cycloid is a tautochrone] to pendulums, we have had to
establish a new theory of curves, namely the theory of curves that are
generated from others by evolution. This leads to comparing the lengths of
curves and straight lines, which I pursued beyond what my subject
required: I did it because of the beauty and apparent novelty of this
theory."[10] The theory of developments of curves, in which second deri-
vatives essentially appeared for the first time in the study of curves, was
one of the first chapters of differential geometry.

We have discussed in detail the cycloid pendulum, to whose invention
Huygens attributed the greatest significance: "[To prove this] it was first of
all necessary to corroborate and amplify the doctrine of the great Galileo
on falling bodies, a doctrine whose most desirable fruit and highest peak,
as it were, is the property of the cycloid that we have discovered."[11]

Centrifugal Force and a Clock with a Conical Pendulum

The cycloid pendulum was not Huygens' only invention in the course of
perfecting the clock. Another direction of his work in chronometry is
associated with the theory of centrifugal force. Huygens created this theory
and it is significant that he first published it in *Horlogium Oscillatorium*. In
the fifth section of this book, he presents, without proof, theorems on
centrifugal force and describes the construction of a clock with a conical
pendulum (it is known that Huygens invented such a clock on October 5,

---

[10] *Oeuvres*, vol. 18, p. 88.
[11] Idem.

1659). The proofs of the theorems are contained in *De Vi Centrifuga* (*On Centrifugal Force*), which was written in 1659 but came to light only eight years after Huygens' death. Aristotle had known about centrifugal force, and Ptolemy assumed that if the earth rotated on its axis then, because of centrifugal force, objects would not remain on its surface. Kepler and Galileo refuted this point of view, explaining that in this case weight counterbalances centrifugal force, essentially proposing that centrifugal force decreases with the distance from the center of rotation. However only Huygens obtained the remarkable formula for centrifugal force, $F_{cf} = mv^2/R$, to which Galileo had come very close. In an appendix we give Huygens' original text and the reader can see in what form (perhaps not the most economical from today's point of view) Huygens first reported his results.

Regardless of what problem Huygens studied, he always thought about possible applications of his results to clocks. Here, too, he wanted to use the conical pendulum. A conical pendulum is a string with a weight that revolves around an axis through its point of suspension. Let l be the length of the string, α be the angle between the string and the vertical, and R be the distance from the weight to the axis. If the pendulum moves in a circle and the angle α remains constant, then $mv^2/R = mg \tan \alpha$. Thus $v = \sqrt{gR \tan \alpha}$. For the period, the time for one revolution, we obtain (since $T = 2\pi R/v$)

$$T = 2\pi \sqrt{\frac{R}{g} \cot \alpha} = 2\pi \sqrt{\frac{l \cos \alpha}{g}} = 2\pi \sqrt{\frac{u}{g}}.$$

Here $u = l \cos \alpha$ is the length of the projection of the string onto the axis of the pendulum.

Huygens' text extensively discusses the formula for the period of a conical pendulum. The motion of a conical pendulum is compared to two motions that had been studied thoroughly by that time: free fall and the oscillations of a simple (or mathematical) pendulum (Huygens called the latter sideways oscillations, as opposed to the circular oscillations of a conical pendulum).

Thus, the period is determined by the projections of the string onto the axis. The difficulty of making an isochronous conical pendulum lies in the fact that its angle with the axis gradually decreases and the period increases. Huygens calculated that for the period to remain constant as the angle decreases, the length of the string must decrease so that its end lies on a paraboloid of revolution.

Indeed, suppose we have a surface of revolution (Huygens took a paraboloid, the surface of revolution of the parabola $py = x^2$ around the y-axis). A point mass revolves stably around a horizontal cross section (a circle) if the sum of the gravitational and centrifugal forces is directed along the normal to the surface (perpendicular to the tangent plane), and thus the formula for a conical pendulum can be applied. In this case, α is the angle of the normal to the axis, l is the length of the section of the normal between the axis and the surface, and u is the projection of this segment onto the axis. The passage here from a conical pendulum to the revolution of a point mass is somewhat analogous to Galileo's passing from a mathematical pendulum to the motion of a point mass along a circular trough. Here Huygens remarks that for the parabola $py = x^2$ the quantity u (the projection of the section of the normal onto the axis) is independent of the location of the point and equals $p/2$. From this he draws the conclusion that the period of revolution of a point mass along any horizontal section of the paraboloid is the same:

$$T = 2\pi \sqrt{\frac{p}{2g}}.$$

This gives a new way to obtain isochronous oscillations which, in Huygens' opinion, was important for making clocks. If we suspend a conical pendulum so that its end moves along the surface of the paraboloid obtained by revolving the parabola $py = x^2$, independent of the angle of inclination α of the string to the axis, then the period of revolution will not depend on α. In other words, we must arrange things so that the length l changes when α changes and guarantee that the projection u onto the axis remains constant. Huygens thought of an extremely clever method of suspension. He proposed to make a plate in the form of a semicubical parabola $y^2 = ax^3 + b$ and to attach one end of the string at some point of the plate. It then turns out that we can choose a, b, and the length of the string so that no matter how we stretch the string and wind part of it onto the plate, its other end will be on the parabola. The secret of this clever method of suspension relies on the same mathematical considerations as the method for supporting a cycloid pendulum.

We note that in 1687 these calculations helped Huygens quickly solve Leibnitz's problem about a curve along which a point mass moves so that the segments it traverses in equal time intervals have equal projections onto the vertical. The semicubical parabola has this property.

The Physical Pendulum

One of Huygens' major achievements involves the theory of the physical pendulum, i.e., not the oscillations of a point mass but those of a configuration of weights or of a plate. This problem arose in connection with the idea of having, besides the basic weight at the end of the pendulum, a moving weight that allowed the period to be regulated. Huygens got this idea from Simon Douw, a craftsman from the Hague, who took out a patent in 1658 on his version of the pendulum clock, differing only slightly from Huygens'. Problems on the oscillations of a physical pendulum had arisen earlier. For mechanics, it was essential to be able to pass from the motion of a point mass to that of extended configurations. The first series of such problems involved the center of gravity, and here important results were known. But for a long time, no real progress had been made on problems about the oscillations of a physical pendulum.[12]

Huygens learned of problems involving the physical pendulum from Mersenne: "When I was still practically a child [less than seventeen—S.G.], the very scholarly Mersenne once suggested to me, as to many others, the study of the center of oscillation or perturbation. That was then a famous problem among the geometers of the time, as I conclude from the letters he wrote to me as well as the recently published writings of Descartes, which contain a response to Mersenne's letters on this subject. . . . At the same time he promised me a great and enviable reward for my work if by chance I managed to satisfy his request. But he did not receive what he wanted from anyone. As for me, as I found nothing that opened the way towards this contemplation but was turned away even at the entrance, I refrained from further study. Those who had hoped to succeed, illustrious men such as Descartes, Honoré Fabry, and others, did not achieve their goal at all, except in the easiest cases, or else they gave, in my opinion, no valid proof. . . . The manner of adjusting the pendulum of our *automate* [clock] by applying in addition to the lower weight, a moveable weight as explained in the description of the clock, gave us an occasion to undertake this research again. Taking up the question under better conditions and from the beginning, I finally triumphed over all difficulties and solved not only Mersenne's problems, but also more difficult ones; I even found at last a general method for finding the center [of oscillations] of lines, surfaces, and solid bodies. From this I had, beyond the pleasure of finding what others had so long searched for and of learning the laws and decrees of nature in these matters, the advantage of knowing henceforth an easy and sure method for adjusting a clock. A second result which seems to me the most important is that I can, based on this theory, give a very precise definition of length, well-defined and invariable over the centuries. . . ."[13]

This last idea of Huygens is that, just as the day is a natural unit for measuring time, a unit for measuring length should be ⅓ the length of a pendulum whose period is one second.

Problems on the center of oscillation were beyond the reach of the methods of mathematical analysis that had been worked out at the time. Huygens noted that a whole series of difficulties could be overcome by beginning with energy considerations: A moving center of gravity cannot

---

[12] Recall that the effective length of a physical pendulum is the length of the mathematical pendulum that oscillates with the same period about a point on the line between the point of suspension and the center of gravity. The distance from the center of the oscillation to the suspension point is the effective length.

[13] *Oeuvres*, vol. 18, pp. 242–244.

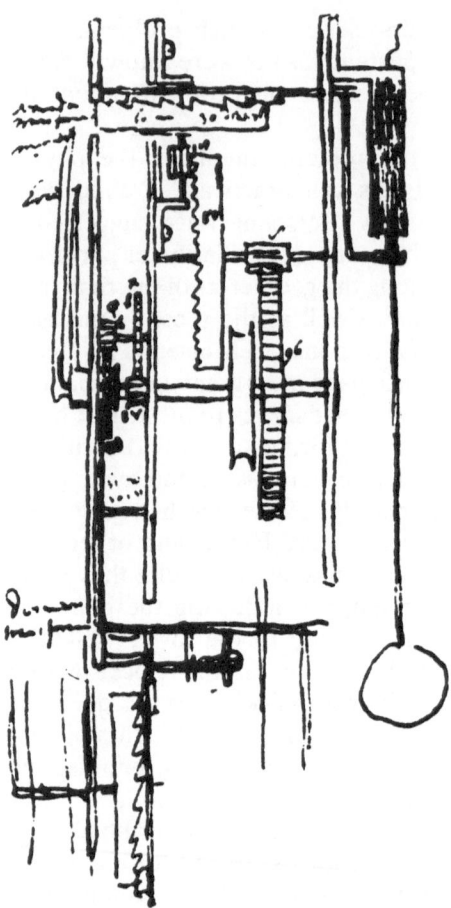

Sketch of a clock with a cycloid pendulum, made by Huygens.

be raised higher than it was at the beginning (otherwise there would be perpetual motion). This method of proof drew comments from many leading scientists, and much effort was spent on it before Jakob Bernoulli succeeded in establishing analogous statements by other means.

Maritime Clocks

The year 1673 was the acme of Huygens' activities on pendulum clocks. *Horlogium Oscillatorium* came out that year, and the Parisian clockmaker Isaac Thuret made a model of a clock that took every improvement into account. Pendulum clocks firmly caught on, but hopes for a maritime pendulum clock were unwarranted. The first models of such clocks had been made in 1661, and sea trials began in 1663. First, Count Alexander Bruce took a clock with him on a voyage from Holland to London, but the clock was slow; Captain Robert Holmes' experiments in sailing from London to Lisbon were more successful. In *Horlogium Oscillatorium*, Huygens tells of the dramatic events associated with clock experiments during the English fleet's voyage to Guinea. Experiments with varying success took place until 1687, although it had become clear that pendulum clocks did not give the hoped-for way of measuring longitude. The demand for a maritime clock gradually subsided, and in 1679 Huygens himself was inclined to think that a spring clock with a balance wheel would have to serve as a maritime clock. In 1735, John Harrison succeeded in making such a chronometer, and received a prize of 20,000 pounds from the English government.

CHRISTIANI

H V G E N I I

ZVLICHEMII, CONST F

HOROLOGIVM

OSCILLATORIVM

SIVE

DE MOTV PENDVLORVM

AD HOROLOGIA APTATO

DEMONSTRATIONES

GEOMETRICA

PARISIIS.

Apud F. Muguet, Regis & Illuftriffimi Archiepifcopi Typographum, vià Citharæ, ad infigne trium Regum.

MDCLXXIII.

CVM TRIVILEGIO REGIS.

Title page from the first edition of *Horlogium Oscillatorium*.

Three hundred years have passed. People have been well served by pendulum clocks, although they have rarely known the name of the man who invented them. The dramatic story of Huygens' work is very instructive. In some sense, his chief ambitions were not realized: He never succeeded in making a maritime chronometer, and the cycloid pendulum, which Huygens considered to be his principal invention, did not survive in land clocks (amplitude restrictors were quite sufficient). The conical pendulum suffered the same fate. But his mathematical and physical results that were motivated by problems on perfecting clocks have lasted to this day in infinitesimal analysis, differential geometry, and mechanics, and one cannot overestimate their significance.

Appendix

Part Five of *Horlogium Oscillatorium*

Containing Another Construction Based on the Circular Motion of Pendulums, and Theorems on Centrifugal Force[14]

...At first, I intended to publish a description of these clocks with the theory of circular motion and centrifugal force—as I wish to call it—a subject on which I had more to say than I had time for at the moment. But so that those interested in these matters could sooner enjoy this new and in no way useless theory and so that publication would not be hindered accidentally, I added this part to the others, against my plans. I briefly describe the construction of the device and at the same time state the relative theorems on centrifugal force, saving their proofs for later.

Construction of the Second Clock

I have not deemed it necessary to set out here the disposition of the gears that form the interior of the clock, since they can easily be arranged to suit the craftsmen and the disposition can be changed in various ways. It suffices to explain the part that controls the movement in a well-determined way. The following figure represents this part.

We must imagine the axis *DH* as being vertical and moveable on two poles. A rather large curved plate is attached to it at *A*; it is curved along *AB*, which is the [semicubical parabola] whose evolution describes a parabola, after a certain line has been adjoined to it, as we have proved in Proposition VIII of Part Three. This line is here *AE*, and the curve *EF* represents the parabola described by the evolution of the entire curve *BAE*. The string applied to *BA* ɛ d whose end describes the parabola is *BCF*. The weight attached to the string is *F*. But, while the axis *DH*

---

[14] Ibid., pp. 360–367.

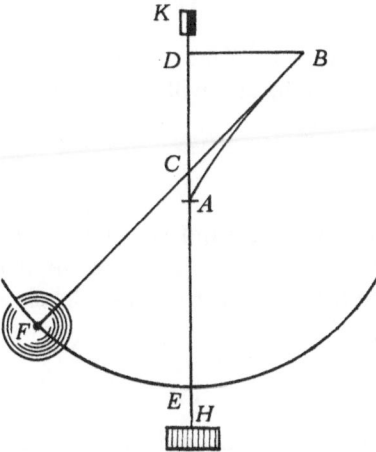

rotates, the stretched string *BCF* makes the ball *F* move so that it travels in horizontal circles, which are larger or smaller according to the force placed on *DH* by the clock gears acting on the small membrane *K*. But these circles lie on the surface of a parabolic conoid, and by this means the revolution times will always be equal, as will follow by what we say below about this motion.

If we wish the clock to show half-seconds, the latus rectum of the parabola *EF* must be 4½ inches of our hour-foot, i.e., half the length of the pendulum whose simple oscillations take half a second. But the length of the latus rectum of the [semicubical parabola] *AB* depends on that of the parabola: The former equals ²⁷⁄₁₆ times the latter. Similarly, the length of *AE* is half that of the latus rectum of the parabola. But if we wish each revolution to last one second, the latera recta and *AE* must be four times as large as before....

Theorems on the Centrifugal Force Resulting from Circular Motion[15]

I

If two identical bodies travel unequal circumferences in equal times, the centrifugal force corresponding to the larger circumference is to that of the smaller as the circumferences themselves or their diameters.

---

[15] Comments on the text are given in square brackets, using the following notation: m is the mass of the body, F the centrifugal force, T the period, R the distance from the center, and v the speed.

II

If two identical bodies move with the same speed around unequal circumferences, their centrifugal forces will be inversely proportional to the diameters.

III

If two identical bodies move around equal circumferences with unequal speeds, each of which is constant, as we assume throughout, then the centrifugal force of the faster one is to that of the slower one as the squares of the speeds.

IV

If two identical bodies moving around unequal circumferences have the same centrifugal force, the time for one revolution around the larger circumference is to that around the smaller as the square roots of the diameters.

V

If a body moves around the circumference of a circle with the speed it acquires in falling from a height of one-fourth the diameter, it will have a centrifugal force equal to its weight, i.e., it will pull the string attaching it to the center with the same force as if it were suspended.

[If the height is $H = R/2$, then the final speed is $v = \sqrt{2gH} = \sqrt{Rg}$ and the centrifugal force is $F = mv^2/R = mRg/R = mg$.]

VI

On the concave surface of a parabolic conoid [paraboloid] with vertical axis, all revolutions of a body moving around horizontal circumferences, small or large, are accomplished in equal times. Each of these times equals that of one double oscillation of a pendulum whose length is half the latus rectum of the generating parabola. . . .

Blaise Pascal

Pascal had his abyss, it followed him.

BAUDELAIRE, *The Abyss*[1]

Blaise Pascal was inherently multifaceted, a characteristic of the Renaissance that had almost become passé in the seventeenth century. The natural sciences (say, physics and mathematics) had not yet completely separated from the humanities, but studies in the humanities and the natural sciences were already no longer commonly combined.

Pascal entered the history of the natural sciences as a great physicist and mathematician, one of the creators of mathematical analysis, projective geometry, probability theory, computational methods, and hydrostatics. France counts him as one of its most remarkable writers: "Narrow minds are surprised by Pascal as the most perfect writer in the greatest century of the French language....Each line coming out of his pen is revered as a precious stone" (Bertrand). Far from everyone agreed with Pascal's thoughts about man, his place in the universe, and the meaning of life, but no one was indifferent to the lines for which the author paid with his life and which have surprisingly not aged. In 1805, Stendahl wrote, "When I read Pascal, it seems to me that I am reading myself." And 100 years later in 1910, Leo Tolstoy read "the wonderous Pascal," "a man of great mind and great heart," and "I could not but be moved to tears, reading him and being conscious of my complete unity with this man who died hundreds of years ago." It is instructive to compare how ideas in the natural sciences and the humanities have aged.

Let us recall one side of Pascal's legacy—his practical achievements. Some were honored most highly, but today few know their author's name. For Turgenev, the standards for convenience and simplicity were "Columbus' egg"[2] and "Pascal's wheelbarrow." Learning that the great scientist had invented a most ordinary wheelbarrow, he wrote to Nekrasov: "Incidentally, in one place I speak of Pascal's wheelbarrow—

---

[1] Charles Baudelaire, *Les Fleurs du Mal* (*The Flowers of Evil*), trans. Richard Howard. Translation copyright © 1983 by Richard Howard. Reprinted by permission of David R. Godine, Publisher, Boston.
[2] this refers to a sixteenth century anecdote about Columbus balancing an egg on its narrow end—*Transl.*

Blaise Pascal, 1623–1662.

you know that Pascal invented this so obviously simple machine." And Pascal also originated the idea of the omnibus, a coach available to everyone ("for 5 sous"), with a fixed route, that was the first form of regular urban transport.

Pascal was one of the most notable people in the history of humanity; an immense literature is devoted to him. What aspects of his life and legacy have not been touched by "Pascalology"? There is peculiar testimony to his special popularity in France: Pascal's portrait is found on French currency (as are portraits of Corneille, Racine, Voltaire, and Pasteur; some time ago notes with Napoleon and Molière were withdrawn from circulation for technical reasons).

Sticks and Coins

When we learn to draw graphs, in the kaleidoscope of anonymous curves we sometimes find ones that are named after people: the spiral of Archimedes, Newton's trident, the conchoid of Nicomedes, the folium of Descartes, the witch of Agnesi, Pascal's limaçon, ... Rarely does anyone doubt that this is the Pascal of "Pascal's law."[3] But this remarkable fourth-degree curve immortalizes the name of Etienne Pascal (1588–1651), Blaise

---

[3] of fluid pressure—*Transl.*

Pascal's father. Etienne Pascal, as was the custom in the Pascal family, served in the Parlement (law court) of the town of Clermont. It was rare to combine legal work with scientific work far from jurisprudence. At about the same time, a councillor at the Parlement of Toulouse, Pierre Fermat (1601–1665), was devoting his leisure time to mathematics. Although Etienne's own achievements were meager, his basic knowledge allowed him to maintain professional contact with most French mathematicians. He exchanged difficult problems on the construction of triangles with the great Fermat, and took Fermat's side in a dispute over maximum and minimum problems with René Descartes (1596–1650). Blaise inherited his father's good relations with many mathematicians, but he also inherited strained relations with Descartes.

Etienne Pascal, an early widower, mostly devoted himself to raising his children (he had two daughters, Gilberte and Jacqueline, besides his son). The young Blaise was soon found to be startlingly gifted but, as often happens, this came along with bad health. (Strange things happened to him all his life; as a young child he almost died from an unknown disease, accompanied by fits that family legend attributed to a witch who had given the child the evil eye.)

Etienne Pascal carefully thought out a system for raising his children. At first he intentionally excluded mathematics from the subjects he taught Blaise: He was afraid that an early enthusiasm for mathematics would interfere with a harmonious development, and that the unavoidable strain of thinking would harm his son's poor health. However the twelve-year-old boy, learning of the existence of a mysterious geometry that his father had studied, convinced him to talk about the forbidden science. The information he received turned out to be enough to begin a fascinating "game with geometry," and to prove theorem after theorem. In this game there were "coins" (circles), "three-cornered hats" (triangles), "tables" (rectangles), and "sticks" (lines). The son was surprised by his father just as he discovered that the angles of a three-cornered hat total the same as two angles of a table. Etienne easily recognized the famous thirty-second proposition of Euclid's first book, the theorem on the sum of the angles of a triangle. The results were tears in the father's eyes and admission to the cabinet that held his mathematics books.

How Pascal constructed Euclidean geometry by himself is known from his sister Gilberte's rhapsodic story. This story created widespread confusion over the notion that since Pascal had discovered the thirty-second proposition of Euclid's *Elements*, he had first discovered all the preceding theorems and axioms. This was not infrequently taken as an argument for Euclid's axioms being the only ones possible. In fact, Pascal's geometry was probably at a "pre-Euclidean" level, where assertions that were not intuitively obvious were proved by reference to obvious ones, and what was obvious was not at all fixed or restricted. It is only at the next, substantially higher, level that the great discovery is made that we can

Pascal in his youth (drawing by Jean Domat).

restrict the obvious assertions to a finite, comparatively small set of axioms which are assumed true, and prove the remaining assertions in geometry from them. Along with proving what is not obvious (e.g., theorems on noteworthy points in triangles), one must also prove the "obvious" theorems that are easy to verify (e.g., the simplest conditions for congruent triangles).

Properly speaking, the thirty-second proposition is the first one in *Elements* that is not obvious in this sense. Without a doubt, the young Pascal had no time to do the enormous job of choosing axioms, let alone any need to do so.

It is interesting to compare this to Einstein's testimony that at the age of twelve he understood geometry, to a significant extent by himself (in particular, he proved the Pythagorean theorem, after hearing about it from his uncle): "It was generally enough for me to base my proofs on those statements whose validity seemed to me indisputable."

At about the age of ten, Pascal did his first work in physics: Interested in the reason for the sound made by a china plate, he carried out a strikingly well organized series of experiments using improvised materials, and explained how the air vibrates.

Hexagramme Mystique, or Pascal's Great Theorem

At thirteen, Blaise Pascal already had access to Mersenne's mathematical circle, which included most of the mathematicians in Paris, including his father Etienne (the Pascals had lived in Paris since 1631).

In the history of science, the French monk Marin Mersenne (1588–1648) played the great and original role of scientist-administrator.[4] His chief merit was that he carried on an extensive correspondence with most of the world's great scientists (he had several hundred correspondents). Mersenne was able to gather information and communicate it to interested scientists. This work required a peculiar gift: the ability to understand new things quickly and to pose questions well. Having great moral character, Mersenne enjoyed the confidence of his correspondents. Sometimes, he wrote to very young scientists. Thus, in 1646 he began to correspond with the seventeen-year-old Huygens, helping him take his first scientific steps and heralding that he would become "the Apollonius and Archimedes... of the coming age."

Together with his "collective" of remote correspondents, there was also a local circle—"Mersenne's Thursdays," into which Blaise Pascal fell. Here he found himself a suitable teacher, Gerard Desargues (1593–1662), an engineer and architect, and creator of an original theory of perspective. His *magnum opus* of 1639, entitled *Brouillon Project d'une Atteinte aux Evénemens des Rencontres du Cône avec un Plan* (*A Proposed Draft of an Attempt to Treat the Results of a Cone Intersecting a Plane*), found few readers. Pascal occupied a special place among them, being able to make considerable strides in this area.

Although at the time Descartes was breaking a completely new trail by creating analytic geometry, geometry in essence had hardly reached the level where it had been in ancient Greece. Much of the legacy of the Greek geometers remained unclear, and this was true most of all for the conic sections. The eight books of Apollonius' *Konika* (*Conic Sections*), the most outstanding work on this theme, were only partially known. Attempts were made to give a modern presentation of the theory, most notably by Claude Mydorge (1585–1647), a member of Mersenne's circle, but his paper really contained no new ideas. Desargues noticed that a systematic application of the method of perspective allowed the construction of a theory of conic sections from a completely new standpoint.

Consider the central projection from a point O above the plane α in the accompanying sketch onto the plane β. It is very natural to apply such a transformation in the theory of conic sections, since their very definitions, as sections of a right circular cone, can be rephrased as follows: All are

---

[4] In evaluating Mersenne's work, we should keep in mind that the first scientific journal, *Journal des Savants*, was founded in 1665.

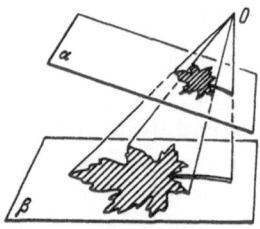

obtained from one by a central projection from the vertex of a cone onto various planes (for example, from a circle). Furthermore, noting that under a central projection intersecting lines can become either intersecting or parallel, we can combine these last two properties into one and assume that all parallel lines meet at one "infinitely distant point." Different bundles of parallel lines give different infinitely distant points, and the infinitely distant points of a plane form an "infinitely distant line." With this understanding, any two distinct lines (including parallel lines) will meet at a unique point. The claim that through any point A not on a line m there is a unique line parallel to m can be reformulated as follows: There is a unique line through an ordinary point A and the infinitely distant point corresponding to the family of lines parallel to m. As a result, under these new hypotheses the following is valid, without any restrictions: There is a unique line through any two distinct points (the line is infinitely distant if both points are infinitely distant). We will see that a very elegant theory results, but it is important for us that under a central projection, the point of intersection of lines (in the generalized sense) is mapped into a point of intersection.

It is important to think about the role played in this assertion by the introduction of infinitely distant elements (under what hypotheses a point of intersection becomes an infinitely distant point, and when a line becomes an infinitely distant line). Without dwelling on the use of this simple idea of Desargues, we will discuss how Pascal applied it so remarkably.

In 1640, Pascal published his *Essai Pour les Coniques (Essay on Conics)*. Here are some facts about this edition that are not devoid of interest: 50 copies were printed and 53 lines of text were printed on posters to be pasted on buildings (it is not known for certain about Pascal, but Desargues was notorious for advertising his results this way). The following theorem, now known as Pascal's theorem, was stated without proof on the poster and signed with the author's initials. *Number six arbitrary points on a conic section L. Let P, Q, R denote the points of intersection of the three pairs of lines (1,2) and (4,5), (2,3) and (5,6), and (3,4) and (6,1). Then P, Q, R are collinear.*[5] In the sketch, L is a

---

[5] The corollary obtained when some of these points are infinitely distant is left as an exercise for the reader.

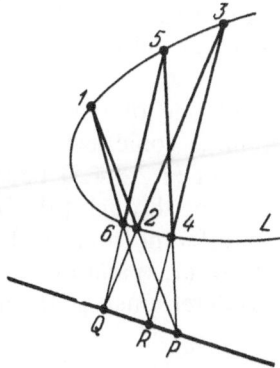

parabola. With the simplest numbering ("in order"), P, Q, and R are the points of intersection of opposite sides of a hexagon.

At first, Pascal stated the theorem for a circle and restricted himself to the simplest numbering of the points. In this case, the problem is elementary but not overly simple. The transition from the circle to an arbitrary conic section is very simple. We must transform such a section to a circle by a central projection, and use the fact that lines are mapped to lines and intersection points (in the generalized sense) are mapped to intersection points. Then, as was already shown, the images of P, Q, and R under a projection will be collinear, and this implies that P, Q, and R themselves are collinear.

This theorem, which Pascal called the theorem on the *hexagramme mystique* (mystic hexagram), was not an end in itself. He considered it the key to constructing a general theory of conic sections, encompassing Apollonius' theory. Generalizations of important theorems of Apollonius, which Desargues did not succeed in obtaining, are mentioned in the poster. Desargues thought a great deal of the result, calling it *la Pascale*. He claimed that it contained the first four books of Apollonius.

Pascal began work on *Traité des Coniques* (*Treatise on Conics*), which he mentioned as being completed in his address *Celeberrimae Matheseos Academiae Parisiensi* (*To the Illustrious Parisian Academy of Science*), in 1654. We know from Mersenne that Pascal obtained about 400 corollaries of this theorem. Gottfried Wilhelm Leibniz (1646–1716) was the last

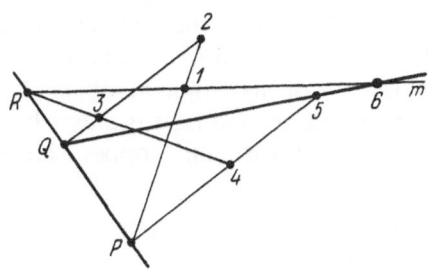

person to see the treatise, after Pascal's death, in 1675–1676. Not heeding Leibniz's advice, Pascal's relatives did not publish the manuscript, and in time it was lost.

As an example, we present one of the simplest but most important corollaries of Pascal's theorem. A conic section is uniquely determined by any five of its points. Indeed, let $\{1,2,3,4,5\}$ be points of a conic section, and m be an arbitrary line passing through (5). Then m contains a unique point (6) of the conic section different from (5). In the notation of Pascal's theorem, P is the intersection of $(1,2)$ and $(4,5)$, Q of $(2,3)$ and m, and R of $(3,4)$ and PQ. Then (6) is determined as the intersection of $(1,R)$ and m.

Pascal's Wheel

On January 2, 1640, Pascal's family moved to Rouen, where Etienne Pascal had obtained a position as *intendant* of the province, and was effectively in charge of all business under the governor.

This appointment heralded fortuitous events. Etienne had taken an active part in the actions of the Parisian investors, for which he was threatened with imprisonment in the Bastille. He had to go into hiding, but at the time Jacqueline came down with smallpox and her father, ignoring the terrible threat to himself, visited her. Jacqueline recovered and even took part in a play attended by Cardinal Richelieu. Thanks to the young actress' appeal, the cardinal pardoned her father but also gave him a job. The former troublemaker had to implement the cardinal's policies (this craftiness will probably not surprise readers of *The Three Musketeers*).

Now Etienne Pascal had much accounting work to do, in which his son regularly helped him. At the end of 1640 Blaise Pascal had the idea of constructing a machine to free his mind from calculations "by counter or pen." The basic idea came quickly and remained unchanged during the course of the work: "...each wheel or pivot of some category, completing a motion of ten numbers, makes the next one move by only one number." However, this was only the first step of a brilliant idea. An incomparably great effort was required to carry it out. In his *avis*,[6] Pascal briefly writes to those who "will have the curiosity to see the arithmetical machine and to make use of it": "I have spared neither time nor trouble nor expense to bring it to a state where it will be useful to you." Before these words came five years of anxious work leading to the creation of a machine ("Pascal's wheel," as his contemporaries called it), which reliably but rather slowly carried out four operations on five-digit numbers. Pascal manufactured about fifty copies of the machine, and here is only a list of the materials he tried: wood, ivory, ebony, brass, and copper. He spent much effort

---

[6] accompanying the dedication of the machine. This appears in Pascal's *Oeuvres Complètes*, pp. 353 ff.—*Transl.*

searching for the best artisans, masters of "the lathe, saw, and hammer," and it often seemed to him that they were unable to achieve the necessary precision. He carefully thought out a system of tests, including a journey of 250 leagues. Pascal did not forget about advertising, either: He enlisted the support of Chancellor Séguier, secured a "royal privilege" (something like a patent), demonstrated his machine often in the salons, and even sent a copy to the Swedish Queen Christina. Finally, he went into production; the exact number of machines produced is unknown, but eight copies still survive.

It is striking how brilliantly Pascal was able to do the most varied things. It became known comparatively recently that in 1623, Schickard, Kepler's friend, built an arithmetical machine, but Pascal's machine was far more perfect.

"Abhorring a Vacuum" and "The Great Experiment on the Equilibrium of Fluids"

At the end of 1646, rumors reached Rouen of surprising "Italian experiments with a vacuum." The question of whether a vacuum can exist in nature had even worried the ancient Greeks. Their opinions on this question revealed the characteristically diverse points of view in ancient Greek philosophy: Epicurus assumed that a vacuum can and does exist; Heron that it can be obtained artificially; Empedocles that there is none and that none cannot be gotten from anywhere; and, finally, Aristotle stated that "nature abhors a vacuum." In the Middle Ages the situation was simpler, since the truth of Aristotle's teaching was practically legislated (even in seventeenth century France, one could be sentenced to hard labor for opposing Aristotle).

The recollection of "abhorring a vacuum" remained for a long time, as the following passage from an unfinished work of Dostoevsky, *Krokodil* (*The Crocodile*), shows: "How is one, in constructing the crocodile, to secure that he should swallow people? The answer is clearer still: construct him hollow. It was settled by physics long ago that Nature abhors a vacuum. Here the inside of the crocodile must be hollow so that it may abhor the vacuum, and consequently swallow and so fill itself with anything it can come across."[7]

Water gives the classical example of "abhorring a vacuum," when it rises in following a piston and does not allow an empty space to form. But suddenly an incident arose over this example. In building the fountains of Florence, it was discovered that water "does not want" to rise more than 34 feet (10.3 meters). The puzzled builders turned for help to the aged Galileo, who joked that nature probably no longer abhors a vacuum above

---

[7] This English translation appears in *An Honest Thief and Other Stories*, trans. Constance Garnett (New York: Macmillan, 1923), p. 277.

34 feet, but proposed anyway that his students Torricelli and Viviani study the strange phenomenon. It was probably Torricelli (and possibly Galileo himself) who thought that the height to which a pump can raise a liquid is inversely proportional to the specific gravity of the liquid. In particular, we should be able to lift mercury 13.3 times less high than water, i.e., to 76 centimeters. This experiment was on a scale more suitable to laboratory conditions, and was conducted by Viviani at Torricelli's initiative. The experiment is well known, but let us recall anyway that a graduated glass tube, sealed at one end, is filled with mercury and the open end closed off with a finger. The tube is inverted and lowered into a cup of mercury. If the finger is removed, the level of mercury in the tube falls to 76 centimeters. Torricelli made two assertions: First, the space above the mercury in the tube is empty (it was later called a "Torricelli vacuum"), and second, the mercury does not completely run out of the tube because it is stopped by the column of air pressing down on the surface of the mercury in the cup. We can explain everything by accepting these hypotheses, but we can also obtain an explanation by introducing special, rather complicated forces that stop a vacuum from forming. It was not a simple matter to adopt Torricelli's hypotheses. Only a few of his contemporaries accepted the idea that air has weight. Some who did believed that a vacuum was possible, but it was almost impossible to believe that air, which is so light, could support the heavy mercury in the tube. Recall that Galileo tried to explain this effect by the properties of the liquid itself, and that Descartes claimed that an apparent vacuum is always filled with "the most fine matter."

Pascal enthusiastically repeated the Italian experiments, thinking of many clever improvements. He describes eight such experiments in a treatise published in 1647. He did not limit himself to mercury, but also experimented with water, oil, and red wine, for which he required barrels instead of cups and tubes about 15 meters long. Spectacular experiments were carried out in the streets of Rouen, to the delight of its inhabitants. (To this day, they like to reproduce engravings with wine barometers in physics textbooks.)

At first, Pascal was most interested in proving that the space above the mercury is empty. It was widely thought an apparent vacuum is filled with matter "that had no properties" (this recalls Lieutenant Kije, from Tinyanov's story, "who had no figure"). It was simply impossible to prove the absence of such matter. Pascal's clear statements are very important in formulating the broader question of the nature of proof in physics. He writes: "Having proved that none of the matter that comes before our senses and of which we know fills this apparently empty space, my feeling will be, until someone has shown me the existence of some matter which fills it, that it is truly empty and devoid of all matter."[8] Less academic

---

[8] From *Expériences Nouvelles Touchant le Vide*, in *Oeuvres*, p. 369.

statements are contained in a letter to the Jesuit scholar Etienne Noël: "But we have more grounds to deny its existence ["the most fine matter"— *S.G.*] because we cannot prove it, than to believe in it for the sole reason that we cannot prove it does not exist."[9] Thus, it is necessary to prove the existence of an object and one can never require a proof of its absence (this is like the legal principle that a court must prove guilt and has no right to require the accused to prove his innocence).

At the time, Pascal's sister Gilberte lived at the family home in Clermont. Her husband, Florin Périer, a court councillor, devoted his free time to science. On November 15, 1647, Pascal sent Périer a letter asking him to compare the levels of mercury in a Torricelli tube at the base and summit of Le Puy de Dôme [a local mountain—*Transl.*]: "If it happens that the height of the quicksilver is less at the top than at the base of the mountain (as I have many reasons to believe it is, although all who have studied the matter are of the opposite opinion), it follows of necessity that the weight and pressure of the air is the sole cause of this suspension of the quicksilver, and not the abhorrence of a vacuum: for it is quite certain that there is much more air that presses on the foot of the mountain than there is on its summit, and one cannot well say that nature abhors a vacuum more at the foot of the mountain than at its summit."[10] The experiment was postponed for various reasons and only took place on September 19, 1648, in the presence of five "people of standing in this town of Clermont." At the end of the year, a brochure appeared containing Pascal's letter and Périer's report, with a scrupulous description of the experiment. The mountain was about 1.5 kilometers high and the difference in the mercury level was 82.5 millimeters. The participants "were so carried away with wonder and delight" by this difference, which Pascal probably did not expect. To make such an estimate beforehand was impossible, since the illusion of the lightness of air was very great. The result was so appreciable that one of the participants, Father De le Mare, had the idea that an experiment on a smaller scale could give similar results. And indeed, the difference in mercury level at the base and at the top of the cathedral of Notre Dame de Clermont, which is 39 meters high, was 4.5 millimeters. If Pascal had admitted this possibility, he would not have waited ten months. Obtaining the news from Périer, he repeated the experiments at the tallest buildings in Paris, obtaining the same results. Pascal called this experiment "the great experiment on the equilibrium of fluids" (this name may cause surprise, since it speaks of the equilibrium of air and mercury, and calls air a fluid). There is one point of confusion in this story: Descartes claimed that he had prompted the idea for the experiment. There was probably

---

[9] *Oeuvres*, pp. 370 ff.
[10] An English translation of this letter appears in *The Physical Treatises of Pascal*, trans. I.H.B. and A.G.H. Spiers (New York: Columbia Univ. Press, 1937), p. 101.

some misunderstanding here, since it is difficult to assume that Pascal consciously denied credit to Descartes.

Pascal continued to experiment, using large siphons along with baro- metric tubes (choosing a short tube so that the siphon did not work). He described the difference in experimental results for various places in France (Paris, Auvergne, and Dieppe). Pascal knew that a barometer could be used as an altimeter, but also understood that the dependence between the level of mercury and the altitude of the location was not simple and had not yet been found. He remarked that the barometric readings at a given place depend on the weather; today the barometer is mostly used for weather forecasting (Torricelli wanted to construct a device for measuring "changes in the air"). Once Pascal decided to compute the total weight of the atmosphere ("For the pleasure of it, I myself made the computation...") He obtained a figure of 8.28×10^{18} French pounds.[11]

We cannot linger over Pascal's other experiments on the equilibrium of liquids and gases that place him, together with Galileo and Simon Stevin (1548–1620), among the founders of classical hydrostatics. These include Pascal's celebrated law, the concept of the hydraulic press, and the substantial development of the principle of virtual displacements. At the same time, he was thinking about, for example, spectacular experiments illustrating Stevin's paradoxical discovery that the pressure of a fluid on the bottom of a vessel depends not on the form of the vessel but only on the level of the fluid. In one experiment, it is obvious at a glance that a 100- pound weight is needed to equalize the pressure on the bottom of one ounce of water. During the course of the experiment the water freezes, and then a one-ounce weight is enough. Pascal showed a distinctive pedago- gical talent. These facts that struck Pascal and his contemporaries are almost as surprising today.

In 1653, Pascal's physics experiments were interrupted by tragic events, which we will discuss below.

"The Geometry of Chance"

In January 1646, Etienne Pascal slipped on the ice and dislocated his hip, almost costing him his life. The reality of losing his father made a terrible impression on Blaise, manifested above all in his health: His headaches became unbearable, he could only move about on crutches, and was only able to swallow a few drops of warm liquid. From the orthopedists who treated his father, Pascal learned of the teachings of Cornelius Jansenius (1585–1638), which were becoming known in France at the time, opposing the Jesuit movement (which had then been in existence for about a

---

[11] Ibid., pp. 65–66. Based on Pascal's description of the value of the *livre* at the time, this equals 3.25×10^{18} kilograms—*Transl.*

hundred years). One incidental aspect of Jansen's teaching made the greatest impression on Pascal: whether the unchecked study of science is permissible, the striving to learn everything, to unravel everything, which is associated most of all with the boundless curiosity of the human mind, or as Jansen wrote, with the "mind's lust." Pascal took his scientific work to be sinful, and his misfortune to be a punishment for that sin. Pascal himself called this event his "first conversion." He resolved to avoid acts that were "sinful and opposed to God." But he did not succeed: We have already gone ahead and we know that he soon devoted every moment that his illness allowed to physics.

His health improved somewhat, and things happened to Pascal that those close to him did not understand well. He courageously bore his father's death in 1651, and his rationalizing, outwardly cold discussion of his father's role in his life sharply contrasted with his reaction five years earlier (he wrote that now his father's presence was not "absolutely necessary," and that he would have only needed it for ten more years, although it would have been helpful all his life).

Then Pascal made some acquaintances that were quite unsuitable for a Jansenist. He traveled in the retinue of the duke of Roannez and met the Chevalier de Méré, a highly educated and intelligent man, but somewhat superficial and self-assured. De Méré's name has come down in history only because his great contemporaries readily associated with him. He contrived to write Pascal letters with lessons on various subjects, not excluding even mathematics. Today all this seems naive and, in the words of Sainte-Beuve, "such a letter is quite enough to ruin a man, its author, in the opinion of posterity." Nevertheless, after a rather protracted time Pascal willingly became friends with de Méré, and he turned out to be the chevalier's capable student in the realm of worldly life.

We now come to the story of how a "problem, posed by a worldly man to a severe Jansenist, became the origin of probability theory" (Poisson). Properly speaking, there were two problems, and as historians of mathematics have explained, both were known long before de Méré. The first question is: How many times should two dice be thrown so that the probability that double six occurs at least once is greater than the probability that it does not occur at all? De Méré solved the problem himself, but unfortunately...by two methods that gave different answers, 24 and 25 throws. Believing the two methods were equally valid, de Méré attacked the "inconstancy" of mathematics. Pascal, believing the correct answer was 25, did not even work out the solution. His major efforts were directed towards solving the second problem, about "the proper division of stakes." We have a game in which all the players (there may be more than two) put their stakes in a "pot." The game is divided into several rounds, and to win the pot a player must win a certain fixed number of rounds. The question is: How should the pot be divided between the players according to the number of rounds they have won, if the game is not played out to the end (no one wins enough rounds to take the whole pot)? In Pascal's words, "de Méré...could not even approach this problem...."

No one in Pascal's circle could understand the solution he proposed, but a suitable interlocutor was found anyway. Between July 29th and October 27th, Pascal exchanged letters with Fermat (via Pierre de Carcavi, who continued Mersenne's work). It is often thought that this correspondence gave birth to the theory of probability. Fermat solved the stakes problem differently from Pascal, and at first a disagreement arose. But in his last letter, Pascal states: "Our common understanding is completely established," and "As I see, the truth is the same in Toulouse and in Paris." He was happy to find a great like-minded person: "From now on, I want to share my thoughts with you as much as possible."

In that same year of 1654, Pascal published one of his most popular works, *Traité du Triangle Arithmetique (Treatise on the Arithmetical Triangle)*. This is now called Pascal's triangle, but it turns out that it was known in ancient India and was rediscovered by Stifel in the sixteenth century. It rests on a simple method for calculating the number of combinations of n objects taken k at a time, $C(n,k)$, by induction on n: $C(n,k) = C(n-1, k) + C(n-1, k-1)$. In this treatise, the principle of mathematical induction was stated for the first time in the form we are accustomed to seeing, although it had in essence been applied earlier.

In 1654, Pascal, in his *Celeberrimae* address to the Parisian Academy, listed the works he was preparing for publication, including a treatise which "may...claim for itself the privilege of having the amazing title of *The Geometry of Chance*."

Louis de Montalte

Soon after her father's death, Jacqueline Pascal entered a convent, and Blaise Pascal missed having someone very close to him. For a time he was attracted by the possibility of living as most people do: He thought about buying a position at court and marrying. But this was not fated to be. In mid-November of 1654, while Pascal was crossing a bridge, the lead pair of horses broke loose and the coach miraculously stopped at the edge of the abyss. From that time, in Lamettrie's words, "in company or at the table, Pascal always needed to be fenced in on his left by chairs or by people, so that he would not see the terrible abyss into which he was afraid of falling, although he knew the price of such an illusion." On November 23rd, he had an unusual attack of nerves. Finding himself in a state of ecstasy, Pascal wrote down on a scrap of paper the thoughts rushing through his head: "God of Abraham, God of Isaac, God of Jacob, but not the god of philosophers and savants...." Later, he transferred the note onto parchment, and after his death both papers were discovered sewn into his doublet. This event is called Pascal's "second conversion."

From that day, according to Jacqueline, Pascal felt a "tremendous disdain for light and an almost insurmountable aversion to everything that belonged to him." He broke off his work and at the beginning of 1655 moved into the monastery of the Port-Royal (a Jansenist stronghold), voluntarily leading a monastic life.

At this time, Pascal wrote *Les Provinciales*[12] (*The Provincial Letters*), one of the greatest works of French literature. *Les Provinciales*, a criticism of the Jesuits, consisted of "letters" published separately from January 23, 1656, to March 23, 1657, eighteen letters in all. The author, a "friend of the provincial," was called Louis de Montalte. The word "mountain" (*la montagne*) in this pseudonym probably recalls the experiments on Le Puy de Dôme. The letters were read throughout France and the Jesuits were enraged, but they could not reply appropriately (the king's confessor, Père Annat, proposed fifteen times, according to letters he wrote at the time, to declare Montalte a heretic). The author, who turned out to be a daring and talented conspirator, was pursued by the judicial investigator, directed by the same Chancellor Séguier who had once supported the creator of the arithmetical machine (according to a contemporary, after just two letters, they had to "bleed the chancellor seven times"), and finally in 1660 the Council of State decided to burn the book of the "imaginary Montalte." But this was essentially a symbolic measure, and Pascal's tactic had striking results. Voltaire wrote about *Provinciales*, "Attempts had been made by the most varied means to show that the Jesuits were abominable; Pascal did more: he showed they were ridiculous." Balzac called them a "chef-d'oeuvre of witty logic," and Racine said they were "buried treasure for comedy." Pascal's works foreshadowed the appearance of Molière's *Tartuffe*.

Working on *Provinciales*, Pascal clearly understood that not only mathematicians need to master logic. Many at the Port-Royal reflected on the educational system, and there were even Jansenist "little schools." Pascal took an active part, for example, making interesting remarks about the first steps towards reading and writing (he believed one should not begin by studying the alphabet). In 1667 two fragments of his work were published posthumously, *De l'Esprit Geométrique et de l'Art de Persuader* (*Geometrical Reasoning and the Art of Persuasion*). These essays do not constitute scientific work; their purpose was more modest, to serve as an introduction to a geometry textbook for the Jansenist schools. Many of Pascal's statements make a very strong impression, and it is hard to believe that such a clear statement was possible in the midseventeenth century. Here is one: "Prove each proposition that is a bit obscure, and in the proof use only use axioms that are quite obvious, or propositions that have been agreed on or proven. Always mentally replace terms that have been defined by their definitions, so as not to be led astray by the ambiguity of

---

[12] also known as *Lettres Provinciales—Transl.*

the terms that the definitions have restricted."[13] Elsewhere, Pascal remarks that there must be undefined concepts. From this, Jacques Hada-
mard (1865–1963) assumed that Pascal was only a small step away from
carrying out a "deep revolution in all of logic—a revolution that Pascal
could have brought about three centuries before it actually occurred."
Here he probably had in mind the view of axiomatic theory that took shape
after the discovery of non-Euclidean geometry.

Surprising events did not cease to take place in Pascal's life. In that terrible year of 1654, his beloved niece Marguerite developed an abcess in the
corner of her eye. The doctors were unable to help the girl, and her condition steadily worsened. In March 1657, a "holy thorn," by legend taken
from Christ's crown of thorns and kept at the Port-Royal, was put into her
eye and the abcess subsided. "The miracle of the holy thorn," in the words
of Gilberte Périer (Marguerite's mother), "was attested to by several sur-
geons and physicians, and authorized by the solemn judgment of the
Church." Rumors about the event made such a strong impression on the
church that the Jansenist monastery in turn escaped being closed. As for
Pascal, she said "his joy was so great that it filled him completely; and as
nothing ever occupied his spirit without much reflection, several very important *pensées* [thoughts] about miracles in general came to him on the
occasion of this particular miracle...."[14] The great scientist believed in
miracles! He wrote[15]: "It is not possible to have a reasonable belief against
miracles." Later, he even tried to define a miracle: "Miracle.—It is an
effect, which exceeds the natural power of the means which are employed
for it...." Many attempts were later made to explain the event rationally
(one explanation was that a metallic speck was the cause of the abcess, and
that the thorn had magnetic properties). From that time on, Pascal's seal
contained the image of an eye surrounded by a crown of thorns.

Amos Dettonville

"I spent a long time in the study of the abstract sciences, and was
disheartened by the small number of fellow-students in them. When I
commenced the study of man, I saw that these abstract sciences are not
suited to man, and that I was wandering farther from my own state in
examining them, than others in not knowing them."[16] These words of

---

[13] *Oeuvres*, p. 597.
[14] From Giberte Périer's *La Vie de Monsieur Pascal* (*The Life of Mr. Pascal*), in
Pascal's *Oeuvres*, p. 15.
[15] These quotations are taken from Pascal's last work, *Pensées* (*Thoughts*), which
appears in English translation by W.F. Trotter (London: Everyman's Library and
J.M. Dent & Sons Ltd).
[16] Ibid., p. 55.

Pascal characterize his mood during the last years of his life. Still, he spent a year and a half of these last years on mathematics....

It began one night during the spring of 1658 when, while suffering from a terrible toothache, Pascal remembered an unsolved problem of Mersenne on the cycloid. He noticed that his intense thinking diverted him from his pain. By morning he had already proven a whole series of results on the cycloid, and...had recovered from his toothache. At first, Pascal felt he had committed a sin and did not intend to write down the results he had obtained. Later, under the influence of the duke of Roannez, he changed his mind; during the course of eight days, according to Gilberte Périer, "as soon as he did something he wrote it down, while his hand could still write." Then in June 1658, Pascal organized a contest, as was often done at the time, in which he posed six problems on cycloids to the best mathematicians. The most successful were Christiaan Huygens, who solved four problems, and John Wallis (1616–1703), who solved all of them, although with some difficulty. But the work that was acknowledged as best belonged to the unknown Amos Dettonville. Huygens later said that "this work was so astutely done that there was nothing to add." Note that "Amos Dettonville" consists of the same letters as "Louis de Montalte." This was Pascal's new pseudonym.[17] Dettonville's work won the prize of 60 *pistoles*.

Now a few words about this work. We have already talked about the cycloid, in the chapter on Huygens. This curve is described by a point on a circle that rolls along a line without slipping. Initial interest in the cycloid was stimulated by the fact that many interesting problems about it could be solved by elementary means. For example, by Torricelli's theorem, in order to construct the tangent to a cycloid at a point A, we must take the position of the generating (rolling) circle corresponding to A and join the highest point B on the circle to A. Another theorem, that Torricelli and Viviani ascribe to Galileo, states: the area of the curvilinear figure bounded by an arc of a cycloid is three times the area of the generating circle.

The problems considered by Pascal no longer had elementary solutions (the area and center of gravity of an arbitrary segment of the cycloid, the volumes of the corresponding solids of revolution, etc.). In these problems, Pascal essentially worked out everything that was needed to construct differential and integral calculus in general form. Leibniz, who shares with Newton the glory of creating this theory, wrote that when, on Huygens' advice, he familiarized himself with Pascal's works, he "was illuminated by a new light." He was surprised to find how close Pascal had

---

[17] Another anagram of this name, Salomon de Tultie, appeared in *Pensées*, among the names of the authors whom he followed (together with Epictetus and Montaigne). Pascalians worked quite a bit to find this mysterious philosopher, until they guessed what was going on.

been to constructing the general theory, and how he unexpectedly stopped short, as if "there were scales before his eyes."

It was characteristic of the works anticipating the appearance of differential and integral calculus that their authors' intuition kept them from being able to produce strict proofs; the language of mathematics was insufficiently developed to put this way of thinking down on paper. A way out was later found, by introducing new ideas and special notation. Pascal did not resort to symbols, but he was such a virtuoso of language that at times it seems that he simply didn't need any. Here is what N. Bourbaki says: "Wallis in 1655 and Pascal in 1658 forged, each for his own use, language of an algebraic nature in which, without writing a single formula, they made statements that can be immediately transcribed into formulas of integral calculus as soon as one understands the technique. Pascal's language is especially clear and precise. And if one does not understand why he refused to use the algebraic notation not only of Descartes but even of Vieta, one can only admire his tour de force, which his mastery of the language rendered him alone capable of accomplishing."[18] We would like to say that here Pascal the writer aided Pascal the mathematician.

Pensées

After mid-1659, Pascal returned to neither physics nor mathematics. At the end of May 1660, he traveled to his native Clermont for the last time; Fermat invited him to come to Toulouse. It is bitter to read Pascal's answer of August 10th. Here are some extracts[19]: "I would also tell you that, although in all Europe you are the one I consider the greatest geometer, it would not be that quality which would draw me; but I imagine such spirit and honesty in your conversation, that it is for that that I would seek you out. . . . I consider [geometry] the highest exercise of the spirit; but at the same time I know it to be so useless that I find little difference between a man who is only a geometer and a skilled artisan. Thus I call it the most beautiful profession in the world; but in the end it is only a profession; and I have often said that it is good for testing our strength but not for using it. . . ." And, finally, here are some lines referring to Pascal's physical condition: "I am so weak that I cannot walk without a stick, nor stay on a horse. I can only go three or four leagues at the most in a carriage. . . ." In December 1660, Huygens twice visited Pascal and found him deeply aged (Pascal was only thirty-seven) and unable to carry on a conversation.

Pascal resolved to look into the most hidden secrets of human existence, in the sense of life. He was perplexed[20]: "I know not who put me into the world, nor what the world is, nor what I myself am. I am in terrible

---

[18] *Éléments d'Histoire des Mathématiques* (Paris: Hermann, 1969), pp. 238–239.
[19] *Oeuvres*, pp. 522–523.
[20] The following quotations are taken from *Pensées*, pp. 21–123, passim.—*Transl.*

ignorance of everything....As I know not whence I come, so I know not whither I go....Such is my state, full of weakness and uncertainty." His studies of the natural sciences cannot help him answer the questions that arise: "Physical science will not console me for the ignorance of morality in the time of affliction." He once wrote, "There are no real proofs anywhere, except in geometry and where it is imitated." But this time geometry could serve as a model (although not a few people have tried to construct a mathematical theory of morals!). Pushkin wrote, not without irony: "'Everything that surpasses geometry surpasses us,' said Pascal. And in consequence of this he wrote his philosophical *pensées*!" But Pascal sees no contradiction here. He searched for the truth elsewhere: "I only approve of those who search with pain in their hearts." He writes: "All our dignity consists, then, in thought. By it we must elevate ourselves, and not by space and time which we cannot fill. Let us endeavour, then, to think well; this is the principle of morality." He returns to this question repeatedly: "Man is obviously made to think. It is his whole dignity and his whole merit; and his whole duty is to think as he ought....Now, of what does the world think?...of dancing, playing the lute, singing, making verses, running at the ring, etc., fighting, making oneself king...." "All the dignity of man consists in thought....But what is this thought? How foolish it is !" But to reflect well is not without danger: "Excess, like defect of intellect, is accused of madness. Nothing is good but mediocrity." Pascal ponders much about the role of religion in human life. There is almost no question that he passes over. He reflects on human history, emphasizing the role of chance ("Cleopatra's nose: had it been shorter, the whole aspect of the world would have been altered"), and speaks of the terrible side of human life ("Can anything be more ridiculous than that a man should have the right to kill me because he lives on the other side of the water, and because his ruler has a quarrel with mine, though I have none with him?"). Pascal's statements on the most diverse questions are extraordinarily astute. His thoughts on government were esteemed by Napoleon, who while imprisoned on the island of St. Helena said that he "would have made Pascal a senator."

Pascal did not complete the major book of his life. The material he left behind was published posthumously in different versions and under different titles. It is most often called *Pensées* (*Thoughts*).

The book was extraordinarily popular, but here we will just stress its influence on the leading figures in Russian culture. Not everyone accepted it. Turgenev called *Pensées* "the most awful, most unbearable book ever published," but wrote that "...never has anyone yet emphasized what Pascal emphasizes: his melancholy, his imprecations are awful. Compared to him Byron is pink lemonade. But what depth, what clarity, what greatness!...Such free, strong, impudent, and mighty language!..." Chernyshevsky wrote about Pascal: "...to perish from an excess of intellectual power—what a glorious death...." Dostoevsky argued with

Pascal all his life. For Tolstoy, Pascal was one of the most revered thinkers. Pascal's name constantly occurs in the *Cycle of Readings* he compiled,[21] about 200 times. For Tolstoy, Pascal is a writer who "writes with his heart's blood."

Blaise Pascal died on August 19, 1662. On August 21st, a burial certificate (*acte d'inhumation*) was drawn up in the church of Saint-Etienne-du-Mont: "On Monday, August 21, 1662, the late Blaise Pascal was buried in this church, in his lifetime Esquire and son of the late Messire Etienne Pascal, councillor of State and president of the *Cour des Aides* of Clermont-Ferrand. Fifty priests. Received: 20 francs."

---

[21] daily readings on truth, life, and behavior, taken from various writers—*Transl.*

Prince of Mathematicians

Nihil actum reputans si quid superesset agendum.
Judging that nothing was done if something was left undone.

GAUSS[1]

In 1854, the health of Privy Councillor Gauss, as his colleagues at the University of Göttingen called him, worsened decisively. There was no question of continuing the daily walks from the observatory to the literary museum, a habit of over twenty years. They managed to convince the professor, who was nearing eighty, to go to the doctor! He improved during the summer and even attended the opening of the Hannover-Göttingen railway. In January 1855, Gauss agreed to pose for a medallion by the artist Hesemann. After the scientist's death in February 1855, a medal was prepared from the medallion, by order of the Hannover court. Beneath a bas-relief of Gauss, these words were written: *Mathematicorum princeps* (Prince of Mathematicians). The story of every real prince should begin with his childhood, embroidered with legends. Gauss is no exception.

I. Gauss' Debut

"The obstinacy with which Gauss followed a path once chosen, the youthful impetuosity with which he regularly and recklessly took the steepest way towards his goal—these hard tests strengthened his powers and made him capable of striding recklessly over all obstacles, even when they had already been removed by earlier investigations. And to this praise of independent activity I would like to add another: the praise of youth. What I want to say perhaps means only that the laws which underlie the development of mathematical genius are the same as those for any other creative gift: in the early years, when a person has just reached full physical growth, great revelations may hurry in upon him; it is then that he creates what he has to bring into the world as his own new value, even though his ability to

---

[1] As translated in Felix Klein, *Development of Mathematics in the 19th Century*, trans. Michael Ackerman (Brookline: Math Sci Press, 1979), p. 8.—*Transl.*

Carl Friedrich Gauss, 1777–1855.

express them may not yet be equal to his abundant flow of ideas" (Felix Klein).[2]

Braunschweig, 1777–1795

Gauss did not inherit his title, although his father Gebhard Dietrich was no stranger to mathematics. A jack-of-all-trades, primarily a fountain builder but also a gardener like his father before him, Gebhard Dietrich was known for his talent as an accountant. Merchants made use of his services during fairs in Braunschweig and even Leipzig, and he was also regularly employed by the largest burial fund in Braunschweig (a position he bequeathed to his son by his first wife, Johann Georg, a retired soldier).

Carl Friedrich was born on April 30, 1777, in house number 1550 on the Wendengraben canal in Braunschweig. According to his biographers, he inherited good health from his father's side and a brilliant mind from his mother's. He was closest to his uncle Johann Friederich Benze, a skillful weaver in whom, in his nephew's words, "an innate genius perished." Gauss said of himself that he "could count before he spoke." The earliest mathematical legend about him claims that at the age of three he followed

---

[2] Ibid., pp. 31–32.

his father's calculations with a bricklayer, unexpectedly corrected him, and turned out to be right.

At seven, Carl Friedrich entered the Catharineum school. Since they only learned to count in the third grade, little Gauss paid no attention for the first two years. Students usually entered the third grade at age ten and studied there until confirmation at fifteen. The teacher, whose name was Büttner, had to work simultaneously with children of different ages and with different preparation. Thus he usually gave some of his students long computations to do so that he could talk to the others. Once a group of students, including Gauss, was asked to sum the integers from 1 to 100. (Different sources name different numbers!) As soon as they were done, the students were to put their slates on the teacher's table, and the order of the slates counted towards their grades. The ten-year-old Gauss put down his slate when Büttner had hardly finished dictating the problem. To everyone's surprise, his answer was even correct. The secret was simple: While the problem was being dictated, Gauss rediscovered the formula for the sum of an arithmetic progression! The child-genius' fame spread throughout little Braunschweig.

In the school where Gauss studied, the teacher's assistant, whose chief duty was to repair the pupils' pens, was a man named Bartels. Bartels was interested in mathematics and owned a few mathematics books. He and Gauss began to study together; they learned about Newton's binomial formula, infinite series, etc.

What a small world it turned out to be! After sometime, Bartels received a chair in pure mathematics at the University of Kazan, and was Lobachevsky's teacher.

In 1788, Gauss entered the gymnasium. Mathematics was not studied there, but rather classical languages were. Gauss happily studied languages and was so successful that he did not even know what he wanted to become —a mathematician or a philologist.

Word of Gauss reached the court, and in 1791, he was presented to Carl Wilhelm Ferdinand, the duke of Braunschweig. The boy lived at court and amused the courtiers with his feats of calculation. Thanks to the duke's patronage, Gauss was admitted to the University of Göttingen in October 1795. At first he attended lectures on philology and almost none on mathematics. But he continued to study mathematics.

Here is a comment by Felix Klein, the noted mathematician who studied Gauss' scientific work at length: "A natural interest, I might even say a certain childlike curiosity, first led the boy to mathematical questions, independently of any outside influence. Indeed, it was simply the art of calculating with numbers that first attracted him. He calculated continually, with overpowering industry and untiring perseverance. By this incessant exercise in manipulating numbers (for example, calculating decimals to an unbelievable number of places) he acquired not only the astounding virtuosity in computational technique that marked him throughout his life, but also an immense memory stock of definite numerical values, and thereby an appreciation and overview of the realm of numbers such as probably no one, before or after him, has possessed. Aside from arithmetic he was occupied with numerical operations on infinite series. From his activity with numbers, and thus in an inductive, 'experimental' way, he arrived quite early at a knowledge of their general relations and laws. . . . It was not so rare in the eighteenth century—for example, with Euler—but stands in sharp contrast to the normal practice of today's mathematicians. . . . All these early intellectual games, devised solely for his own pleasure, were first steps towards a great goal that became conscious only later. It is part of the anticipatory wisdom of genius to place the pick-ax precisely on the rock vein where the gold mine lies concealed, and to do this even in the half-playful first testings of its powers, unconscious of its deeper meaning. We now come to the year 1795, of which we have more detailed evidence. . . . Then, still before his Göttingen period, a passionate interest in the integers seized him, even more tenaciously than before, as is vividly evidenced by the preface to the *Disquisitiones Arithmeticae*. Unacquainted with the literature, he had to create everything for himself. Here again it was the untiring calculator who blazed the way into the unknown. Gauss set out huge tables: of prime numbers, of quadratic residues and non-residues, and of the fractions $1/p$ for $p = 1$ to 1000 with their decimal expansions carried out to a complete period, and therefore sometimes to several hundred places! With this last table Gauss tried to determine the dependence of the period on the denominator p. What researcher of today would be likely to enter upon this strange path in search of a new theorem? But for Gauss it was precisely this path, followed with such unheard of energy—he himself maintained that he differed from other men only in his diligence—that led to his goal. . . . In the autumn of 1795 he moved to Göttingen, where he must have devoured the works of Euler and Lagrange, presented to him for the first time."[3]

A Discovery after Two Thousand Years

On June 1, 1796, the following notice appeared in the newspaper *Jenenser Intelligenzblatt*: "Every beginner in geometry knows that it is possible to construct different regular polygons [with compass and straightedge], for example triangles, pentagons, 15-gons, and those regular polygons that result from doubling the number of sides of these figures. One had already come this far in Euclid's time, and it seems that since then one has generally believed that the field for elementary geometry ended at that point, and in any case I do not know of any successful attempt to extend the boundaries beyond that line.

---

[3] Ibid., pp. 29–30.

Therefore it seems to me that this discovery possesses special interest, *that besides these regular polygons, a number of others are geometrically constructible, for example the 17-gon.*"[4]

Beneath the notice was the signature, *C.F. Gauss, Braunschweig*, Mathematics Student at Göttingen.

This is the first communication of a discovery by Gauss. Before discussing it in detail, let us refresh our memories about what "every beginner in geometry knows."

Constructions with Straightedge and Compass

Suppose we are given an interval of length 1. Using a straightedge and compass, we can construct new intervals whose lengths are obtained from those we already have by addition, subtraction, multiplication, division, and extracting square roots.

By successively carrying out these operations, we can construct, using straightedge and compass, any interval whose length can be expressed in terms of 1 by a finite number of these operations. We call these numbers quadratic irrationals. One can prove that these are the only intervals that can be constructed with straightedge and compass.

It is easy to see that the problem of constructing a regular n-gon is equivalent to the problem of dividing a circle of radius 1 into n equal parts. The chords of the arcs into which we divide the circle are the sides of a regular n-gon, and their lengths are all equal to $2 \sin (\pi/n)$. Thus *for each n for which $\sin(\pi/n)$ is a quadratic irrational, we can construct a regular n-gon with straightedge and compass. This condition is satisfied, for example, by $n = 3, 4, 5, 6,$ and 10.* This is well known for $n = 3, 4,$ and 6.

We will show that $\sin (\pi/10)$ is a quadratic irrational. Consider an isosceles triangle ABC whose vertex angle B equals $\pi/5 = 36°$ and whose base AB has length 1. Let AD bisect angle A. Then $x = AC = AD = BD = 2 \sin (\pi/10)$. We have

$$\frac{BD}{DC} = \frac{AB}{AC}; \qquad \frac{x}{1 - x} = \frac{1}{x}, \qquad x = \frac{\sqrt{5} - 1}{2}.$$

This is a quadratic irrational number, so we can construct the sides of a regular 10-gon.

Furthermore, if we can divide a circle into $p_1 p_2$ equal parts then we can of course divide it into p_1 equal parts (in particular, we can construct a regular pentagon). The converse is in general not true, but we note two special cases when it does hold.

1) If we can divide a circle into p equal parts, then we can divide it into $2^k p$

---

[4] This appears in English in Tord Hall, *Carl Friedrich Gauss*, trans. Albert Froderberg, copyright © 1970 by The M.I.T. Press, p. 24.—*Transl.*

equal parts, for any k. This is true because we can bisect any angle with straightedge and compass.

2) If we can divide a circle into p_1 equal parts and into p_2 equal parts, and p_1 and p_2 are relatively prime (e.g., if p_1 and p_2 are distinct primes), then we can divide the circle into $p_1 p_2$ equal parts. This follows from the fact that the greatest common divisor of the angles $2\pi/p_1$ and $2\pi/p_2$ is $2\pi/p_1 p_2$, and the greatest common divisor of two commensurable angles can be found by straightedge and compass. In particular, $2\pi/15 = \frac{1}{2}(2\pi/3 - 2\pi/5)$, which implies the possibility of constructing a regular 15-gon.

A Few Words about Complex Numbers

We need to know just a bit about complex numbers: their basic operations and geometric interpretation. Recall that a complex number $z = a + ib$ corresponds to a point with coordinates (a, b) and to the vector from the origin $(0, 0)$ to this point. The length of the vector, $r = \sqrt{a^2 + b^2}$ is called the *modulus* of the given number z, and is denoted by $|z|$. We can write z in trigonometric form: $z = a + ib = r(\cos\varphi + i\sin\varphi)$. The angle φ is called the *argument* of z.

Addition of complex numbers corresponds to vector addition. In multiplying two complex numbers, we multiply their moduli and add their arguments. This imples that the equation $z^n = 1$ has exactly n roots, which are usually denoted by

$$\varepsilon_k = \cos\left(\frac{2\pi k}{n}\right) + i\sin\left(\frac{2\pi k}{n}\right), \qquad k = 0, 1, \ldots, n-1. \tag{1}$$

As vectors, the ε_k end at the vertices of a regular n-gon. If we prove that the ε_k are quadratic irrationals (i.e., that their real and imaginary parts a and b are quadratic irrationals), then this will prove that the regular n-gon can be constructed with straightedge and compass.

Regular n-Gons and Roots of Unity

We rewrite $z^n = 1$ as

$$z^n - 1 = (z - 1)(z^{n-1} + z^{n-2} + \ldots + z + 1) = 0.$$

We obtain two equations, $z = 1$ and

$$z^{n-1} + z^{n-2} + \ldots + z + 1 = 0. \tag{2}$$

The roots of equation (2) are the numbers ε_k, for $1 \le k \le n - 1$, and we will work with this equation below.

For $n = 3$, we obtain the equation $z^2 + z + 1 = 0$, with roots

$$\varepsilon_1 = -\frac{1}{2} + i\frac{\sqrt{3}}{2}, \qquad \varepsilon_2 = -\frac{1}{2} - i\frac{\sqrt{3}}{2}.$$

For $n = 5$ the situation is more complicated, since we have the fourth-degree equation

$$z^4 + z^3 + z^2 + z + 1 = 0, \tag{3}$$

with four roots ε_1, ε_2, ε_3, and ε_4. Although we have Ferrari's formula for solving the general fourth-degree equation, in practice it is impossible to use. In our case there is a special form of equation (3) that is useful. In order to solve (3), we first divide by z^2, obtaining

$$z^2 + \frac{1}{z^2} + z + \frac{1}{z} + 1 = 0, \quad \text{or} \quad \left(z + \frac{1}{z}\right)^2 + \left(z + \frac{1}{z}\right) - 1 = 0.$$

We make the substitution $w = z + 1/z$:

$$w^2 + w - 1 = 0, \tag{4}$$

with roots

$$w_{1,2} = \frac{-1 \pm \sqrt{5}}{2}.$$

We can also find ε_k from the equations

$$z + \frac{1}{z} = w_1, \qquad z + \frac{1}{z} = w_2, \tag{5}$$

but this is unnecessary. To find them it is enough to know that twice the real part of ε_1 equals

$$2\cos\left(\frac{2\pi}{5}\right) = \varepsilon_1 + \varepsilon_4 = \varepsilon_1 + \frac{1}{\varepsilon_1} = w_1 = \frac{-1 + \sqrt{5}}{2}.$$

Since w_1 is a quadratic irrational, so are ε_1 and ε_4. The argument for ε_2 and ε_3 is similar.

Thus, for $n = 5$ the solution to our problem can be reduced to the successive solution of two quadratic equations: first solve (4), whose roots are the sums $\varepsilon_1 + \varepsilon_4$ and $\varepsilon_2 + \varepsilon_3$ of the symmetric roots of (3), and then find these roots of (3) from (5).

This is the very method Gauss used to construct the regular 17-gon, singling out groups of roots whose sums are found successively from quadratic equations. But how can we look for these "good" groups? Gauss found a surprising way to answer this question.

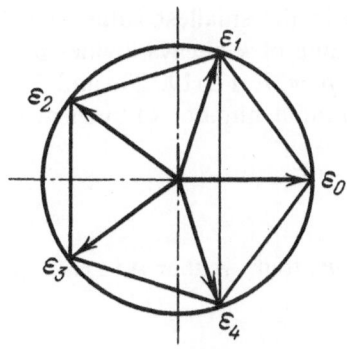

Constructing the Regular 17-Gon

"And on March 30, 1796 he underwent his conversion on the road to Damascus....For a long time Gauss had been busy with grouping the roots of unity $x^n = 1$ on the basis of his theory of 'primitive roots.' Then suddenly one morning, still in bed, he saw clearly that the construction of the 17-gon follows from his theory. As already mentioned, this discovery marked a turning point in Gauss' life. He decided to devote himself entirely to mathematics, not philology" (Felix Klein).[5]

Let us go into a bit more detail about the path that Gauss took. One of the young Gauss' mathematical games was to divide 1 by various primes p and write down the successive decimal digits, impatiently waiting for them to begin to repeat. Sometimes he had a long wait. For $p = 97$ the repetition began with the 97th digit, and for $p = 337$ with the 337th. Gauss was not confused by long lines of computations, but rather entered the mysterious world of numbers with their help. He was not too lazy to consider all $p < 1000$ (see the earlier quote from Klein).

It is known that Gauss did not immediately try to prove the periodicity of these fractions in general ($p \neq 2, 5$), but the proof probably gave him no difficulty. Indeed, it is enough just to note that we must keep track of the remainder rather than the quotient. The digits of the quotient begin to repeat when at the preceding step the remainder equals 1 (why?). This means that we must find some k such that $10^k - 1$ is divisible by p. Since there are only a finite number of possible remainders (they lie between 1 and $p - 1$), for some $k_1 > k_2$ the numbers 10^{k_1} and 10^{k_2} leave the same remainder after dividing by p. But then

$$10^{k_1 - k_2} - 1 \text{ is divisible by } p$$

(why?).

It is a little harder to show that we can always take $p - 1$ as k, i.e., that $10^{p-1} - 1$ is always divisible by p, when $p \neq 2, 5$. This is a special case of the result known as Fermat's Little Theorem. When Fermat (1601–1655) discovered it, he wrote that he "was illuminated by a clear light." Now the young Gauss had rediscovered it. He would always value this result highly: "This theorem is worthy of attention both because of its elegance and its great usefulness."[6]

Gauss was interested in the smallest value of k for which $10^k - 1$ is divisible by p. Such a value of k always divides $p - 1$, and sometimes it equals $p - 1$, e.g., for $p = 7, 17, 19, 23$, and 29. It remains unknown whether there are an infinite number of such primes p or only a finite number.

---

[5] *Development*, p. 30.

[6] *Disquisitiones Arithmeticae*, trans. Arthur A. Clarke (New Haven: Yale Univ. Press, 1966), p. 32.

Gauss replaced 10 by any number a for which p does not divide a, and asked when $a^k - 1$ is not divisible by p for $k < p - 1$. In this case we say a is a *primitive root* mod p. This is equivalent to saying that all the nonzero remainders $1, 2, \ldots, p - 1$ occur when we divide $1, a, a^2, \ldots, a^{p-2}$ by p (why?).

Gauss did not know at the time that Euler (1707–1783) had been interested in primitive roots. Euler conjectured (but could not prove) that *each prime number has at least one primitive root*. Legendre (1752–1833) gave the first proof of Euler's conjecture, and Gauss gave a very elegant proof. But this was later, and for the time being Gauss worked with concrete examples. He knew, for instance, that $a = 3$ *is a primitive root* mod $p = 17$. In the first row of the table below we show the values of k, and below them the remainders of 3^k after dividing by 17. Note that the second row contains all the remainders from 1 to 16, which implies that 3 is primitive mod 17.

0	1	2	3	4	5	6	7	8	9	10	11	12	13	14	15
1	3	9	10	13	5	15	11	16	14	8	7	4	12	2	6

These calculations lay at the heart of grouping the roots of

$$z^{16} + z^{15} + z^{14} + \ldots + z + 1 = 0 \tag{6}$$

in order to reduce its solution to a chain of quadratic equations. Gauss' idea was to renumber the roots. We will give a root ε_k the number s and denote it by $\varepsilon_{[s]}$ if 3^s divided by 17 has remainder k. We can use the table to go from one numbering to the other, finding k in the second row and the corresponding s above it, but it is more convenient to use the accompanying sketch, where the old numbers appear outside the circle and the new ones inside. This is the numbering Gauss used to divide the roots of equation (6) into groups and reduce its solution to a chain of quadratic equations.

At the first stage, let $\sigma_{2,0}$ and $\sigma_{2,1}$ be the sums of the roots $\varepsilon_{[s]}$ for even and odd s, respectively. Each is the sum of the eight roots for which

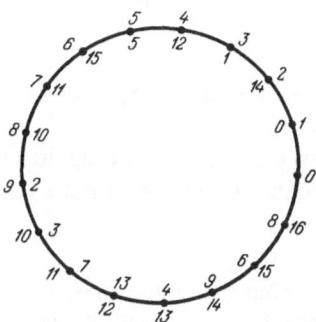

dividing s by 2 leaves the remainder 0 and 1, respectively. These sums turn out to be the roots of a quadratic equation with integer coefficients. Next, let $\sigma_{4,0}$, $\sigma_{4,1}$, $\sigma_{4,2}$, and $\sigma_{4,3}$ be the sums of the four roots $\varepsilon_{[s]}$ for which dividing s by 4 leaves the remainder 0,1,2, and 3, respectively. We will show that these quantities are the roots of quadratic equations whose coefficients can be expressed arithmetically in terms of $\sigma_{2,0}$ and $\sigma_{2,1}$. Finally, form the sums $\sigma_{8,i}$ of the two roots $\varepsilon_{[s]}$ for which dividing s by 8 leaves the remainder i. We write quadratic equations for them whose coefficients are expressed in a simple way in terms of the $\sigma_{4,j}$. We have $\sigma_{8,0} = 2 \cos(2\pi/17)$, and the quadratic irrationality of $\sigma_{8,0}$ implies the possibility of constructing the regular 17-gon with straightedge and compass. It is instructive to write down the decomposition of the roots in the old numbering. You should agree that it is impossible to guess the decomposition in this form. We will now carry out the scheme we have just described.

Detailed Calculations

We will now prove that the seventeenth-order roots of unity are quadratic irrationals. Note that $\varepsilon_k \varepsilon_s = \varepsilon_{k+s}$ (if $k + s \geq 17$, then replace $k + s$ by its remainder after division by 17) and $\varepsilon_k = (\varepsilon_1)^k$. We first remark that

$$\varepsilon_1 + \varepsilon_2 + \cdots + \varepsilon_{16} = \varepsilon_{[0]} + \varepsilon_{[1]} + \cdots + \varepsilon_{[15]} = -1.$$

(For example, consider this as the sum of a geometric progression.)

Let $\sigma_{m,r}$ denote the sum of $\varepsilon_{[k]}$ for those k leaving remainder r after division by m. We obtain

$$\sigma_{2,0} = \varepsilon_{[0]} + \varepsilon_{[2]} + \varepsilon_{[4]} + \cdots + \varepsilon_{[14]},$$

$$\sigma_{2,1} = \varepsilon_{[1]} + \varepsilon_{[3]} + \varepsilon_{[5]} + \cdots + \varepsilon_{[15]}.$$

Clearly,

$$\sigma_{2,0} + \sigma_{2,1} = \varepsilon_{[0]} + \varepsilon_{[1]} + \cdots + \varepsilon_{[15]} = -1.$$

One can show that

$$\sigma_{2,0} \cdot \sigma_{2,1} = 4(\varepsilon_{[0]} + \varepsilon_{[1]} + \cdots + \varepsilon_{[15]}) = -4.\,^{[7]}$$

Now, by Vieta's theorem, we can construct a quadratic equation with roots $\sigma_{2,0}$ and $\sigma_{2,1}$:

$$x^2 + x - 4 = 0, \qquad x_{1,2} = \frac{-1 \pm \sqrt{17}}{2}.$$

In order to distinguish $\sigma_{2,0}$ from $\sigma_{2,1}$, we again use the above sketch. In each sum, the roots appear together with their conjugates. It is clear that

---

[7] We can see this by carrying out the multiplication directly, using $\varepsilon_k \varepsilon_s = \varepsilon_{k+s}$ and the preceding sketch. However, we will learn below a method for avoiding these tedious calculations.

$\sigma_{2,0} > \sigma_{2,1}$, since $\sigma_{2,0}$ is twice the sum of the real parts of ε_1, ε_2, ε_4, and ε_8, while for $\sigma_{2,1}$ we repeat this with ε_3, ε_5, ε_6, and ε_7. Thus

$$\sigma_{2,0} = \frac{\sqrt{17} - 1}{2}, \qquad \sigma_{2,1} = \frac{-\sqrt{17} - 1}{2}.$$

Consider the roots taken four at a time:

$$\sigma_{4,0} = \varepsilon_{[0]} + \varepsilon_{[4]} + \varepsilon_{[8]} + \varepsilon_{[12]},$$

$$\sigma_{4,1} = \varepsilon_{[1]} + \varepsilon_{[5]} + \varepsilon_{[9]} + \varepsilon_{[13]},$$

$$\sigma_{4,2} = \varepsilon_{[2]} + \varepsilon_{[6]} + \varepsilon_{[10]} + \varepsilon_{[14]},$$

$$\sigma_{4,3} = \varepsilon_{[3]} + \varepsilon_{[7]} + \varepsilon_{[11]} + \varepsilon_{[15]}.$$

We have $\sigma_{4,0} + \sigma_{4,2} = \sigma_{2,0}$ and $\sigma_{4,1} + \sigma_{4,3} = \sigma_{2,1}$. One can also show that $\sigma_{4,0} \cdot \sigma_{4,2} = \sigma_{2,0} + \sigma_{2,1} = -1$, which means that $\sigma_{4,0}$ and $\sigma_{4,2}$ are roots of $x^2 - \sigma_{2,0} x - 1 = 0$. Solving this equation and noting that $\sigma_{4,0} > \sigma_{4,2}$ (see the above sketch), after some simple transformations we obtain

$$\sigma_{4,0} = \frac{1}{4}(\sqrt{17} - 1 + \sqrt{34 - 2\sqrt{17}}),$$

$$\sigma_{4,2} = \frac{1}{4}(\sqrt{17} - 1 - \sqrt{34 - 2\sqrt{17}}).$$

Analogously,

$$\sigma_{4,1} = \frac{1}{4}(-\sqrt{17} - 1 + \sqrt{34 + 2\sqrt{17}}),$$

$$\sigma_{4,3} = \frac{1}{4}(-\sqrt{17} - 1 - \sqrt{34 + 2\sqrt{17}}).$$

We now go to the last step. Set

$$\sigma_{8,0} = \varepsilon_{[0]} + \varepsilon_{[8]} = \varepsilon_1 + \varepsilon_{16},$$

$$\sigma_{8,4} = \varepsilon_{[4]} + \varepsilon_{[12]} = \varepsilon_4 + \varepsilon_{13}.$$

We could have considered the six other such expressions but we do not need to, since it suffices to prove that $\sigma_{8,0} = 2\cos(2\pi/17)$ is a quadratic irrational, thus permitting the construction of the regular 17-gon. We have $\sigma_{8,0} + \sigma_{8,4} = \sigma_{4,0}$ and $\sigma_{8,0} \cdot \sigma_{8,4} = \sigma_{4,1}$. The sketch shows that $\sigma_{8,0} > \sigma_{8,4}$, so $\sigma_{8,0}$ is the largest root of $x^2 - \sigma_{4,0} x + \sigma_{4,1} = 0$, i.e.,

$$\sigma_{8,0} = 2\cos\frac{2\pi}{17} = \frac{1}{2}(\sigma_{4,0} + \sqrt{(\sigma_{4,0})^2 - 4\sigma_{4,1}})$$

$$= \frac{1}{8}(\sqrt{17} - 1 + \sqrt{34 - 2\sqrt{17}})$$

$$+ \frac{1}{4}\sqrt{17 + 3\sqrt{17} - \sqrt{170 + 38\sqrt{17}}}.$$

We have transformed somewhat the expression we obtain directly for

$$\sqrt{(\sigma_{4,0})^2 - 4\sigma_{4,1}},$$

but we will not tire the reader by reproducing these simple calculations.

Using this formula for $\cos(2\pi/17)$, we can complete the construction of the regular 17-gon using elementary rules for constructing quadratic irrational expressions. This is a rather awkward procedure, and these days rather compact construction methods are known. We will present one in an appendix, without proof. But the formula for $\cos(2\pi/17)$ leaves no doubt that it would have been impossible to attain within the confines of the traditional geometric ideas of Euclid's time. Gauss' solution belonged to another mathematical epoch. Note that the most interesting claim is the possibility in principle of constructing the regular 17-gon; the procedure itself is not essential. To prove the construction is possible it is enough to see that at each step we have quadratic equations whose coefficients are quadratic irrationals, without writing down explicit expressions for them (this becomes particularly essential when the number of sides is larger).

In the solution we have described for equation (6), we have left completely unexplained why the numbering $\varepsilon_{[s]}$ for decomposing the roots turned out to be successful. How could we have guessed it from the beginning? We will now essentially repeat the solution, exposing the key idea—symmetry in the set of roots.

Symmetry in the Set of Roots of Equation (6)

First of all, the problem of the roots of unity is closely connected to the arithmetic of remainders after division by n (arithmetic modulo n). Indeed, if $\varepsilon^n = 1$, then ε^k is also an nth root of unity and its value depends only on the remainder left after dividing k by n. Set $\varepsilon = \varepsilon_1$ (see formula (1)); then $\varepsilon_k = \varepsilon^k$, so $\varepsilon_k \cdot \varepsilon_s = \varepsilon_{k+s}$, where the sum is taken modulo n (the remainder after dividing by n). In particular, $\varepsilon_k \cdot \varepsilon_{n-k} = \varepsilon_0 = 1$.

Problem 1. If p is prime and δ is any complex pth root of unity, then the powers δ^k, $k = 0, 1, \ldots, p - 1$, contain all the pth roots of unity.

Remark. We must prove here that for each $0 < m < p$, the numbers $0, 1, \ldots, p - 1$ are all contained among the remainders after dividing km by p, for $k = 0, 1, \ldots, p - 1$.

We denote exponentiation to the kth power by T_k: $T_k \varepsilon_s = (\varepsilon_s)^k = \varepsilon_{sk}$.

Problem 2. Prove that if $n = p$ is prime, then each transformation T_k, $k = 1, 2, \ldots, p - 1$, is a one-to-one transformation of the set of roots onto itself, i.e., the set $\{T_k \varepsilon_0, T_k \varepsilon_1, \ldots, T_k \varepsilon_{p-1}\}$ coincides with the set of roots $\{\varepsilon_0, \varepsilon_1, \ldots, \varepsilon_{p-1}\}$.

Problem 1 shows that for each $1 \leq s \leq p - 1$, $\{T_0 \varepsilon_s, T_1 \varepsilon_s, \ldots, T_{p-1} \varepsilon_s\}$ is the set of all roots. Problems 1 and 2 imply the following conclusion: *construct a table whose entry in the kth row and sth column is $T_k \varepsilon_s$, for $1 \leq$

$k, s \leq p - 1$. *Then each row and each column contain all the roots* $\varepsilon_1, \varepsilon_2,$
$\ldots, \varepsilon_{p-1}$ *in some order, without repetition.* Note that $T_{p-1}\varepsilon_s = \varepsilon_{-s} = (\varepsilon_s)^{-1}$.
Those who know the definition of a group may verify that the transformations T_k form a group with respect to the multiplication $T_k \cdot T_s = T_{ks}$.

Now consider the case $p = 17$. We call a set M of roots invariant with respect to T_k if $T_k\varepsilon_s$ belongs to M for each ε_s *in* M. The only set that is invariant with respect to all the T_k is the set $\{\varepsilon_1, \ldots, \varepsilon_{16}\}$ of all roots.

The underlying conjecture is that *a group of roots is "better" when more transformations leave it invariant.*

We introduce another numbering $T_{[s]}$ for T_k, as we did for ε_k: $T_{[s]} = T_k$ when $k = 3^s$. In the new notation,

$$T_{[k]}\varepsilon_{[s]} = \varepsilon_{[k+s]},$$

$$T_{[m]}(T_{[k]}\varepsilon_{[s]}) = T_{[m+k]}\varepsilon_{[s]},$$

where the sum in brackets is taken modulo 16. The reader will of course discover an analogy with logarithms, which is not surprising since

$$\varepsilon_{[s]} = \varepsilon_{3^s}.$$

Problem 3. Prove that if some set of roots is invariant with respect to some $T_{[k]}$, where k is odd, then this set is invariant with respect to all $T_{[m]}$, i.e., if it is nonempty then it is the set of all roots.

Remark. It suffices to show that if k is odd, then there is some m such that 16 divides km with remainder 1.

On the other hand, there are two groups of roots that are invariant with respect to $T_{[k]}$ for all even k: the roots $\varepsilon_{[s]}$ for even s and for odd s. We denote their sums by $\sigma_{2,0}$ and $\sigma_{2,1}$.

Clearly, $\sigma_{2,0} + \sigma_{2,1} = -1$. Consider $\sigma_{2,0} \cdot \sigma_{2,1}$. This is the sum of pairwise products $\varepsilon_{[k]} \cdot \varepsilon_{[s]}$, where k is even and s is odd, each of which is a root $\varepsilon_{[m]}$—64 terms in all. We will show that each root $\varepsilon_{[0]}, \varepsilon_{[1]}, \ldots,$ $\varepsilon_{[15]}$ occurs among them exactly four times, so $\sigma_{2,0} \cdot \sigma_{2,1} = -4$. We will use the fact that $T_{[k]}$ preserves groups of roots when k is even and transforms one into the other when k is odd. Each term of $\sigma_{2,0} \cdot \sigma_{2,1}$ can be uniquely represented in the form $\varepsilon_{[m]}\varepsilon_{[m+r]}$, where $0 \leq m \leq 15$ and $r = 1, 3, 5, 7$ (proof!). Group together terms with the same value of r. We obtain a sum of the form

$$\varepsilon_{[0]}\varepsilon_{[r]} + \varepsilon_{[1]}\varepsilon_{[r+1]} + \cdots + \varepsilon_{[15]}\varepsilon_{[r+15]}$$
$$= T_{[0]}(\varepsilon_{[0]}\varepsilon_{[r]}) + T_{[1]}(\varepsilon_{[0]}\varepsilon_{[r]}) + \cdots + T_{[15]}(\varepsilon_{[0]}\varepsilon_{[r]})$$
$$= T_{[0]}\varepsilon_{[r]} + T_{[1]}\varepsilon_{[r]} + \cdots + T_{[15]}\varepsilon_{[r]}$$
$$= \varepsilon_{[0]} + \varepsilon_{[1]} + \cdots + \varepsilon_{[15]} = -1.$$

We have used the fact that

$$T_{[m]}\varepsilon_{[k]} \cdot T_{[m]}\varepsilon_{[s]} = T_{[m]}(\varepsilon_{[k]}\varepsilon_{[s]}),$$

and the properties of $T_{[m]}$ we have mentioned above.

The values of $\sigma_{2,0}$ and $\sigma_{2,1}$ were found above.

We now go to the next step. We want to introduce new, smaller, groups of roots that are invariant with respect to some $T_{[k]}$. By analogy with problem 3, one can show that k must be divisible by 4. Therefore there are four groups of roots, invariant with respect to all $T_{[4s]}$ and smaller than the ones already considered. Let the sums of the roots in each group be $\sigma_{4,0}$, $\sigma_{4,1}$, $\sigma_{4,2}, \sigma_{4,3}$. We have already noted that $\sigma_{4,0} + \sigma_{4,2} = \sigma_{2,0}$ and $\sigma_{4,1} + \sigma_{4,3} = \sigma_{2,1}$.

We calculate the product $\sigma_{4,0} \cdot \sigma_{4,2}$; it is the sum of sixteen terms of the form $\varepsilon_{[4k]}\varepsilon_{[4s + 2]}$. Each term can be written uniquely in the form $\varepsilon_{[2m]} \cdot \varepsilon_{[2m + 2r]}$, where $r = 1, 3$ and $m = 0, 1, 2, 3, 4, 5, 6, 7$. We group together terms with the same r and note that $\varepsilon_{[0]}\varepsilon_{[2]} = \varepsilon_1\varepsilon_9 = \varepsilon_{10} = \varepsilon_{[3]}$ and $\varepsilon_{[0]}\varepsilon_{[6]} = \varepsilon_1\varepsilon_{15} = \varepsilon_{16} = \varepsilon_{[8]}$. For $r = 1$, we obtain the sum

$$T_{[0]}\varepsilon_{[3]} + T_{[2]}\varepsilon_{[3]} + \cdots + T_{[14]}\varepsilon_{[3]} = \sigma_{2,1};$$

and for $r = 3$, the sum $\Sigma_k T_{[2k]}\varepsilon_{[8]} = \sigma_{2,0}$, i.e., $\sigma_{4,0} \cdot \sigma_{4,2} = \sigma_{2,0} + \sigma_{2,1} = -1$. By solving the quadratic equations, we found $\sigma_{4,0}$ and $\sigma_{4,2}$.

In the last step we consider groups of roots invariant with respect to $T_{[8]}$; there are eight. In particular, $\sigma_{8,0} + \sigma_{8,4} = \sigma_{4,0}$. We compute $\sigma_{8,0} \cdot \sigma_{8,4}$. Taking into account $\varepsilon_{[0]} \cdot \varepsilon_{[4]} = \varepsilon_1\varepsilon_{13} = \varepsilon_{14} = \varepsilon_{[9]}$, we obtain $\sigma_{8,0} \cdot \sigma_{8,4} = T_{[0]}\varepsilon_{[9]} + T_{[4]}\varepsilon_{[9]} + T_{[8]}\varepsilon_{[9]} + T_{[12]}\varepsilon_{[9]} = \sigma_{4,1}$. This allowed us to find $\sigma_{8,0} = 2 \cos(2\pi/17)$ and thus to complete the solution.

We have seen that Gauss' argument is completely based on transformations that rearrange the roots. Lagrange (1736–1813) was the first to consider the role of such transformations in the solvability of equations. Gauss was probably not familiar at the time with Lagrange's work. Later, Galois (1811–1832) placed the study of these transformations at the foundation of the remarkable theory that now bears his name. Essentially, Gauss constructed Galois theory full-blown for the cyclotomic equation of the circle.

Possible Generalizations and Fermat Primes

If we do not try to obtain an explicit expression for the roots but only try to prove they are quadratic irrationals, then we can almost entirely omit the calculations, and consider only the idea of invariance. Namely, $\sigma_{2,0} \cdot \sigma_{2,1}$ is the sum of certain roots $\varepsilon_{[s]}$, and since this sum is transformed into itself by the action of each transformation $T_{[k]}$, all the roots it contains occur the same number of times. Thus $\sigma_{2,0} \cdot \sigma_{2,1}$ is an integer. Analogously, $\sigma_{4,0} \cdot \sigma_{4,2}$ does not change under all transformations of the form $T_{[2k]}$ and is thus a combination of the $\sigma_{2,j}$; $\sigma_{8,0} \cdot \sigma_{8,4}$ is preserved by all $T_{[4k]}$ and is thus a combination of the $\sigma_{4,j}$.

This abbreviated argument allows us to establish those primes p to which we can generalize Gauss' proof to show that the pth roots of unity are quadratic irrationals. An analysis shows that we have only used the fact that $p - 1 = 2^k$ (at each step the groups were divided in half), and the numbering of the roots, which relied on the primitivity of 3 modulo the prime 17. We could have used any primitive root for the numbering. As we have already noted, every prime p has at least one primitive root (incidentally, one can show that 3 is primitive for all p of the form $2^k + 1$ [proof!]). We also remark that if $p = 2^k + 1$ is prime, then $k = 2^r$. Thus, *we have proved it is possible to construct a regular p-gon with straightedge and compass for all primes p of the form*

$$p = 2^{(2^r)} + 1.$$

Primes of this form have their own history, and are called *Fermat primes*. Fermat proposed that all such numbers are prime. Indeed, for $r = 0$ we obtain 3, for $r = 1$ we have 5, and for $r = 2$ we get 17. For $r = 3$ we obtain 257 and $r = 4$ yields 65,537, both of which are prime. For $r = 5$ we obtain 4,294,967,297. Fermat found no prime divisors of this number, but Euler explained that Fermat had "overlooked" the divisor 641. We now know that the Fermat numbers are composite for $r = 6, 7, 8, 9, 11, 12, 15, 18, 23, 36, 38, 73$ (for example, $5 \cdot 2^{75} + 1$ is a prime divisor of $r = 73$). It has been conjectured that there are only a finite number of Fermat primes.

As for regular n-gons with composite n, the properties mentioned above (p. 120) immediately imply that the desired construction is possible for all $n > 2$ of the form $n = 2^k p_1 p_2 \ldots p_s$, where $p_1, p_2, \ldots, p_s$ are distinct Fermat primes. Remarkably, *there are no other values of n for which the construction is possible*. Gauss did not publish a proof of this claim: "The limits of the present work exclude this demonstration here, but we issue this warning lest anyone attempt to achieve geometric constructions [i.e., with straightedge and compass] for sections other than the ones suggested by our theory (e.g., sections into 7, 11, 13, 19, etc., parts) and so spend his time uselessly."[8] Gauss' result implies the possibility in principle of constructing a regular p-gon for $p = 257$ and 65,537, but calculating the roots, let alone describing the construction explicitly, requires a colossal but completely automatic effort. It is remarkable that people were found who wanted to carry this out not only for $p = 257$ (Richelot did it in an 80-page paper, and there is reason to believe that Gauss himself also did) but also for $p = 65,537$ (the solution obtained by Hermes is in Göttingen, in a trunk of considerable proportions). The English mathematician John Littlewood once joked about this: "A too-persistent research student drove his supervisor to say, 'Go away and work out the construction for a regular polygon of 65,537 $[= 2^{16} + 1]$ sides.' The student returned twenty years later with a construction (deposited in the Archives of Göttingen)."[9]

---

[8] *Disquisitiones*, p. 459.
[9] *A Mathematician's Miscellany* (London: Methuen & Co., 1953), p. 42.

Concluding Remarks

We have already noted that March 30, 1796, the day Gauss found the construction of the regular 17-gon, decided his fate. Felix Klein writes:

"On this date begins the diary....We see the proud series of great discoveries in arithmetic, algebra and analysis parade before us.... Among these traces of the burgeoning of a mighty genius one finds, touchingly, little miniatures of school exercises, which even a Gauss was not spared. Here we find a record of conscientious exercises in differentiation; and just before a section on the division of the lemniscate, there are totally banal integral substitutions, such as every student must practice."[10]

Gauss' work has long stood as an unattainable model of mathematical discovery. One of the founders of non-Euclidean geometry, Bolyai János[11] (1802–1860), called it "the most brilliant discovery of our time and even of all time." But it was difficult to comprehend! Thanks to letters to his homeland written by the great Norwegian mathematician Abel (1802–1829), who proved that the general fifth-degree equation is not solvable in radicals, we know the difficult path he followed in studying Gauss' theory. In 1825, Abel wrote from Germany: "Even if Gauss is the greatest genius, he evidently did not try to have everything understood all at once." He decided not to meet with Gauss, but later wrote from France: "I finally succeeded in lifting the veil of mystery that has so far surrounded the theory of division of the circle, created by Gauss." Gauss' work inspired Abel to construct a theory in which "there are so many remarkable theorems that it is simply unbelievable." He then planned to go to Germany, to "take Gauss by storm." Gauss also undoubtedly influenced Galois.

All his life, Gauss retained a touching love for his first discovery: "It is said that Archimedes willed that a monument be placed over his grave in the form of a sphere and cylinder, in memory of his having found that the ratio of the volumes of a cylinder and a sphere inscribed in it is 3:2. Like Archimedes, Gauss expressed the wish that the 17-gon be immortalized in a monument on his grave. This shows what significance Gauss himself placed on his discovery. This picture is not on Gauss' tombstone, but a monument erected to Gauss in Braunschweig stands on a seventeen-sided pedestal, although this is hardly noticeable to the observer" (H. Weber).

Addendum

Here is an extract from Coxeter's *Introduction to Geometry*,[12] containing Richmond's recipe for constructing the regular 17-gon:

---

[10] *Development*, p. 30.
[11] also known as Johann Bolyai—*Transl.*
[12] H.S.M. Coxeter, *Introduction to Geometry*, copyright © 1969 by John Wiley & Sons, Inc., p. 27. Reprinted by permission of John Wiley & Sons, Inc.

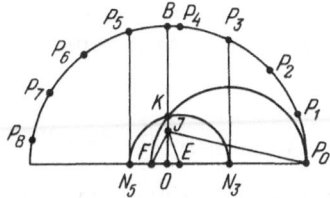

Join P_0 to J, one quarter of the way from O to B. On the diameter through P_0 take E, F, so that $\angle OJE$ is one quarter of OJP_0 and $\angle FJE$ is $45°$. Let the circle with diameter FP_0 cut OB in K, and let the circle with center E and radius EK cut OP_0 in N_3 (between O and P_0) and N_5. Draw perpendiculars to OP_0 at these two points, to cut the original circle in P_3 and P_5. Then the arc P_3P_5 (and likewise P_1P_3) is $\frac{2}{17}$ of the circumference. (The proof involves repeated application of the principle that the roots of the equation $x^2 + 2x \cot 2C - 1 = 0$ are $\tan C$ and $-\cot C$.)

II. The Golden Theorem

...I chanced on an extraordinary arithmetic truth...since I considered it so beautiful in itself and since I suspected its connection with even more profound results, I concentrated on it all my efforts in order to understand the principles on which it depended and to obtain a rigorous proof. When I succeeded in this I was so attracted by these questions that I could not let them be.

GAUSS[1]

Gauss' diary, the chronicle of his remarkable discoveries, begins on March 30, 1796, the day he constructed the regular 17-gon. The next entry appears on April 8th, and talks of the proof of what he called the *theorema aureum* (golden theorem). Fermat, Euler, and Lagrange had proved special cases of this assertion. Euler stated the general hypothesis, and Legendre gave an incomplete proof. On April 8th, Gauss found a complete proof of Euler's hypothesis. Incidentally, Gauss did not yet know of the work of his great predecessors. He had traveled the difficult road to the *theorema aureum* independently!

It all began with childish observations. Sometimes, looking at a very large integer, we can immediately say that its square root is not an integer. For example, we can use the fact that squares of integers cannot end in 2, 3, 7, or 8. Sometimes we can use the fact that the square of an integer is either divisible by 3 or has remainder 1, but never 2. Both properties are of the same type, since the last digit is the remainder after division by 10.

---

[1] *Disquisitiones Arithmeticae*, p. xviii.

Gauss was interested in the general problem: *what are the possible remainders when we divide a square by a prime?* We will now consider this question.

Quadratic Residues

We will assume below that p is a prime number not equal to 2. When we divide an integer by p, we may have a "deficit" or a "surplus," i.e., we can consider either positive or negative remainders. *We will agree to choose the remainder that is smallest in absolute value.*

It is not hard to prove that if p is odd, then every integer n can be written uniquely as

$$n = pq + r, \qquad |r| \leq \frac{p-1}{2}, \tag{1}$$

where q and r are integers.

We will call r the remainder after dividing n by p, or the *residue of n modulo p*, denoted as:

$$n \equiv r \pmod{p}.^2$$

Let us write the residues for the first few primes $p > 2$ in Table 1. We are interested in the possible residues (remainders) of squares of integers. We will call these *quadratic residues*, and the others quadratic nonresidues.

The numbers n^2 and r^2, where r is the remainder of n modulo p, have the same remainder after division by p. Therefore, if we want to find quadratic residues, it suffices to square only residues, i.e., integers r with $|r| \leq k = \frac{1}{2}(p-1)$. It is enough to consider $r \geq 0$.

We carry out the calculations for the primes in Table 1 and construct a new table (Table 2) in which the boldface numbers are quadratic residues.

---

[2] What we are calling the residue (remainder) is usually called the smallest absolute residue (remainder). We have shortened its name, since we will not be dealing with other residues. The notation $n \equiv r \pmod{p}$ is also used to mean that p divides $n - r$.

TABLE 1

p	$k = \dfrac{p-1}{2}$	Residues (remainders) according to size
3	1	$-\ 1\ 0\ 1$
5	2	$-\ 2\ -\ 1\ 0\ 1\ 2$
7	3	$-\ 3\ -\ 2\ -\ 1\ 0\ 1\ 2\ 3$
11	5	$-\ 5\ -\ 4\ -\ 3\ -\ 2\ -\ 1\ 0\ 1\ 2\ 3\ 4\ 5$
13	6	$-\ 6\ -\ 5\ -\ 4\ -\ 3\ -\ 2\ -\ 1\ 0\ 1\ 2\ 3\ 4\ 5\ 6$
17	8	$-\ 8\ -\ 7\ -\ 6\ -\ 5\ -\ 4\ -\ 3\ -\ 2\ -\ 1\ 0\ 1\ 2\ 3\ 4\ 5\ 6\ 7\ 8$

TABLE 2

p	k	Quadratic residues and nonresidues by size
3	1	$-\,1\,0\,1$
5	2	$-\,2\,-\,1\,0\,1\,2$
7	3	$-\,3\,-\,2\,-\,1\,0\,1\,2\,3$
11	5	$-\,5\,-\,4\,-\,3\,-\,2\,-\,1\,0\,1\,2\,3\,4\,5$
13	6	$-\,6\,-\,5\,-\,4\,-\,3\,-\,2\,-\,1\,0\,1\,2\,3\,4\,5\,6$
17	8	$-\,8\,-\,7\,-\,6\,-\,5\,-\,4\,-\,3\,-\,2\,-\,1\,0\,1\,2\,3\,4\,5\,6\,7\,8$

Let us try to find some patterns and see how general they are. First, *in each row there are exactly $k + 1$ boldface numbers.* We will show that this holds for all primes $p > 2$. It follows from what we said earlier that each odd p (nonprime as well) has at most $k + 1$ quadratic residues. We can show there are exactly $k + 1$ if we see that the numbers r^2, for $0 \le r \le k$, have distinct remainders after division by p. If $r_1 > r_2$ and $r_1^2\, r_2^2$ have the same remainder, then p divides $r_1^2 - r_2^2$. Since p is prime, it divides either $r_1 + r_2$ or $r_1 - r_2$, which is impossible since $0 < r_1 \pm r_2 < 2k < p$. This is the first time we have used the fact that p is prime (show our claim is not true for composite numbers).

Fermat's Theorem and Euler's Criterion

Furthermore, 0 *and* 1 *are clearly boldface in each row.* No pattern is immediately visible for the boldface numbers in the other columns. Begin with $a = -1$. It is bold for $p = 5, 13, 17, \ldots$, but not for $p = 3, 7, 11, \ldots$. You may have noticed that the primes in the first group have remainder 1 after division by 4, while those in the second have -1 (for primes $p \ne 2$, no other remainders are possible). Thus we can propose that -1 *is a quadratic residue for primes of the form $p = 4s + 1$ and a quadratic nonresidue for $p = 4s - 1$.* This pattern was first noted by Fermat, but he left no proof. Try to prove it! You will see that the main difficulty lies in finding how to use the assumption that p is prime. It is not at all clear how to do this, and without this assumption the claim becomes false.

After several unsuccessful attempts, the first proof was found by Euler in 1747. In 1755 he found a different, quite elegant proof, using *Fermat's Little Theorem: If p is prime, then for each integer a, $0 < |a| < p$,*

$$a^{p-1} \equiv 1 \pmod{p}. \qquad (2)$$

Proof. For $p = 2$ the assertion is obvious, and so we may assume p is odd. Consider the p numbers $0, \pm a, \pm 2a, \pm 3a, \ldots, \pm ka$, where $k = \frac{1}{2}(p - 1)$. These have distinct remainders after division by p, since otherwise p divides $r_1 a - r_2 a$ for some $r_1 > r_2$, $|r_1| \le k$, $|r_2| \le k$. But p divides neither a nor $r_1 - r_2$, since $0 < r_1 - r_2 < p$. Multiply these numbers together, except for 0, to get $(-1)^k (k!)^2 a^{p-1}$. Since all nonzero residues are among the remainders of these factors, and by the rule for the remainder of a product, we find that the product has the same residue as $(-1)^k(k!)^2$, i.e., p divides $(k!)^2(a^{p-1} - 1)$. Since p does not divide $k!$ $(0 < k < p)$, it divides $a^{p-1} - 1$, which completes the proof.

Corollary (Euler's criterion for quadratic residues). *A residue $b \neq 0$ is quadratic if and only if*

$$b^k \equiv 1(\mathrm{mod}\ p), \ k = \frac{p-1}{2}. \tag{3}$$

Proof. It is easy to establish that condition (3) is necessary. If $a^2 \equiv b$ (mod p), $0 < a < p$, then $a^{2k} = a^{p-1}$ and b^k must each have residue equal to 1, by (2). It is more complicated to prove sufficiency, and we will deduce it from the following lemma.

Lemma 1. *Suppose $P(x)$ is a polynomial of degree s and p is prime. If there are more than s distinct residues r modulo p for which*

$$P(r) \equiv 0 \ (\mathrm{mod}\ p), \tag{4}$$

then (4) holds for all residues.

Proof. We proceed by induction on s. For $s = 0$ the assertion is obvious. Suppose it is true for all polynomials of degree at most $s - 1$. Let $r_0, r_1, \ldots, r_s$, $0 \leq r_j \leq p$, satisfy $P(r) \equiv 0$ (mod p). We represent $P(x)$ in the form $P(x) = (x - r_0) Q(x) + P(r_0)$, where $Q(x)$ is a polynomial of degree $s - 1$ and p divides $P(r_0)$. Then p divides $(r_j - r_0) Q(r_j)$ for $1 \leq j \leq s$. Since p cannot divide $r_j - r_0$ it must divide $Q(r_j)$, and so by the induction assumption p divides $Q(r)$ for all r. Thus, p divides $P(r)$ for all r.

We will apply the lemma to $P(x) = x^k - 1$. Then the k nonzero quadratic residues satisfy (4). But there is a residue $(r = 0)$ not satisfying (4), so by the lemma no quadratic nonresidue can satisfy (4), and thus (3) is also sufficient.

Remark. If b is a quadratic nonresidue, then $b^{(p-1)/2} \equiv -1$ (mod p). Indeed, if $b^{(p-1)/2} \equiv r$ (mod p), then $r^2 \equiv 1$ (mod p), so $r = -1$. (Only the residues $r \equiv 1$ (mod p) and $r \equiv -1$ (mod p) satisfy the congruence $r^2 \equiv 1$ (mod p).)

Euler's criterion allows us to decide instantly for which p the residue -1 is quadratic. Substituting $b = -1$ into (3), we obtain that (3) holds when $p = 4s + 1$ (k is even) but not when $p = 4s - 1$ (k is odd). The hypothesis we stated above is now a theorem.

Problem 1. Prove that if $p \neq 2$ is a prime divisor of $n^2 + 1$, then $p = 4s + 1$.

Thus we have proved that -1 *is a quadratic residue for* $p = 4s + 1$ *and a quadratic nonresidue for* $p = 4s - 1$.

Let us consider several aspects of this proof. The assertion consists of two parts: a negative assertion for $p = 4s - 1$ and a positive one for $p = 4s + 1$. For the former, it is natural to try to find some property that quadratic residues satisfy but -1 does not, which is what Euler did. The property we found turned out to be characteristic, i.e., we proved the latter at the same time. If you try to prove this independently, you would probably try to construct a specific number n^2 whose remainder after division by $p = 4s + 1$ is -1. Euler's proof is not effective in the sense that it does not explicitly construct n in terms of p, but only confirms its existence. In other words, it guarantees that if we go through the numbers $1, 2, \ldots, 2s$, divide their squares by p and take the remainders, then sooner or later we will get -1. The question remains whether there is a more explicit construction of n and p that does not use this exhaustive procedure. In 1773, Lagrange (1736–1813) gave a positive answer, using the following theorem.

Wilson's Theorem.[3] *If* $p = 2k + 1$ *is prime, then*

$$(-1)^k (k!)^2 \equiv -1 \pmod{p}. \tag{5}$$

We use Lemma 1 to prove this theorem. Set $P(x)$ be the product $(x^2 - 1)(x^2 - 4)\ldots(x^2 - k^2)$ and $Q(x) = x^{2k-1} - 1$. Then $R(x) = P(x) - Q(x)$ is a polynomial of degree at most $2k - 1$, which is divisible by p for $x = \pm 1, \pm 2, \ldots \pm k$ (since this is true for P and Q). By the lemma, $R(x) \equiv 0 \pmod{p}$ for all x. The only really new fact is that $R(0) \equiv 0 \pmod{p}$. Since $R(0) = (-1)^k (k!)^2 + 1$, we obtain (5).

Lagrange's Corollary. For $p = 4s + 1$, $[(2s)!]^2 \equiv -1 \pmod{p}$.

Problem 2. Prove that if (5) is true, then p is prime.

This problem gives us an excuse to note that in Lagrange's construction, it is essential that p be prime.

Having explained when $a = -1$ is a quadratic residue, Euler, using an enormous amount of data, tried to find analogous conditions for other values of a. He noticed that for $a = 2$ everything depends on the remainder after dividing p by 8. In fact, 2 turns out to be a quadratic residue for primes $p = 8s \pm 1$ and a nonresidue for $p = 8s \pm 3$ (the remainder after dividing an odd prime by 8 can only be ± 1 and ± 3). Moreover, 3 is a quadratic residue for $p = 12s \pm 1$ and a quadratic nonresidue for $p =$

---

[3] John Wilson (1741–1793) was a jurist who studied mathematics at Cambridge.

$12s \pm 5$. Euler conjectured that in general, everything is determined by the remainder after dividing p by $4a$.

Euler's Hypothesis.[4] Let $0 < r < 4a$. Either a is a quadratic residue for all primes in the arithmetic progression $4aq + r$, $q = 0, 1, 2, \ldots$, or it is a quadratic residue for none.

Clearly, if $4a$ and r have a common divisor $t > 1$, then there will be no primes in this progression. If they are relatively prime, then, by a theorem of Dirichlet (1805–1859), the progression contains infinitely many primes (this generalizes the theorem that there are infinitely many primes in the sequence of all natural numbers).

Let us return to Euler's hypothesis. It turned out that his criterion, which served us well for $a = -1$, spurns us even for $a = 2$. Euler was unable to work through this case. Apart from $a = -1$, he was only able to prove his hypothesis for $a = 3$. Lagrange, whom we have already mentioned, later proved it for $a = 2,5,7$. In 1785 Legendre proposed a general proof, but it contained gaps in essential places.

Gauss' Proof. Gauss, like his predecessors, first treated the case $a = -1$ and then, already divining the general result, examined case after case and advanced further than the others: He considered $a = \pm2, \pm3, \pm5$, and ±7. The general case (Euler's hypothesis) did not yield to the first attack: "This theorem bothered me for a whole year and did not yield to the most strenuous efforts." Note that this was where Gauss "caught up to" the mathematics of his time: The efforts of the best mathematicians, who were trying to prove Euler's hypothesis, were fruitless.

Finally, on April 8, 1796, he found a general proof, which Kronecker (1823–1891) quite aptly called "a test of the power of Gauss' genius." The proof was by double induction on a and p. Gauss had to conceive eight essentially different arguments for eight different cases! He must have been not only strikingly inventive but also surprisingly courageous to continue on this path. Gauss later found six other proofs of the *theorema aureum* (about fifty are now known). As often happens, once the theorem was proved many simpler proofs were found. We will present here a proof that differs only a little from Gauss' third proof. At its heart lies a key lemma, proved by Gauss no earlier than 1808.

Lemma 2. *Let $p = 2k + 1$ be prime, a be an integer, $0 < |a| \leq 2k$, $r_1, r_2, \ldots,$ r_k be the residues of a, $2a, \ldots, ka$, and υ be the number of negative residues among them. Then*

---

[4] This is what Gauss called *theorema aureum*.

$$a^k \equiv (-1)^v \pmod{p}. \tag{6}$$

Applying Euler's criterion, we obtain the following corollary:

Gauss' Criterion for Quadratic Residues. *A residue is quadratic if and only if v is even.*

Proof of Lemma 2. Note that the absolute values of the residues $r_1, r_2, \ldots, r_k$ are all distinct. This follows from the fact that p does not divide the sum or difference of any two of them: $r_i \pm r_j = (i \pm j)a$, $i \neq j$, $|i \pm j| < p$, $|a| < p$. Thus, the moduli $|r_1|, \ldots, |r_k|$ are the numbers $1, \ldots, k$ in some order. So $a \cdot 2a \ldots \cdot ka = a^k k!$ has the same remainder after division by p as $r_1 \ldots r_k = (-1)^v k!$. Since the prime p does not divide $k!$, we obtain (6).

Proof of Euler's Hypothesis. We note that the fact that p is prime is no longer used in the argument, but it was used in full measure in Gauss' lemma. If $a > 0$, we mark off the points $mp/2$ on a number line, while if $a < 0$ we mark off $-mp/2$, where $m = 0, 1, 2, \ldots, |a|$ (Figs. 1a,b). They divide the line into intervals, which we number according to their left endpoints. Now we mark the points $a, 2a, \ldots, ka$ with crosses. Since a is an integer not divisible by p, the crosses cannot fall at the points we marked but fall within the intervals ($|a|p/2 > |a|k$). It is easy to see that the number v in the lemma is *the number of crosses in the odd-numbered intervals* (prove it!).

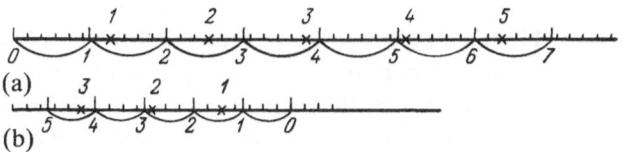

FIGURE 1. (a) $p = 11$ $(k = 5)$, $a = 7$, $v = 3$; (b) $p = 7$ $(k = 3)$, $a = -5$, $v = 2$.

We now apply a similarity transformation with coefficient $1/a$ to our picture (Fig. 1 becomes Fig. 2). The points $mp/2$ are mapped to points that divide $[0, p/2]$ into $|a|$ equal parts, and the crosses are mapped to the integers $1, 2, \ldots, k$.

The numbering of the intervals will now depend on the sign of a. They are numbered according to their left endpoints for $a > 0$, and to their right for $a < 0$; v is the number of integers in the odd-numbered intervals. If we increase p by $4as$, then we add exactly $2s$ integer points to each interval. This follows because translating an interval by an integer does not change the number of integers it contains, and every interval of integer length n with noninteger endpoints contains exactly n integers (prove it!). Thus, in passing from p to $p + 4as$ the value of v changes by an even number, and $(-1)^v$ does not change. This means that $(-1)^v$ is the same for all p in the

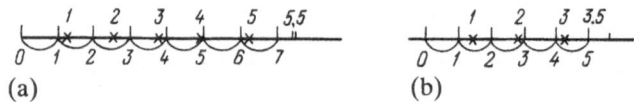

FIGURE 2.

arithmetic progression $p = 4aq + r$, which proves Euler's hypothesis.

This simultaneously gives a method for deciding if a is a quadratic residue for p. Divide p by $4a$ and let r be the remainder (positive for convenience). Divide $(0, r/2)$ into $|a|$ parts, numbered according to their left (right) endpoints if a is positive (negative). Count the number υ of integers in the odd-numbered intervals. Then a is a quadratic residue if and only if υ is even.

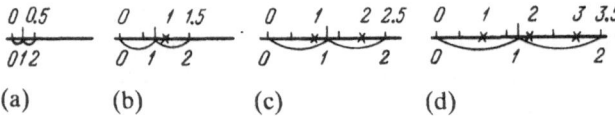

FIGURE 3. (a) $r = 1$, $a = 2$, $\upsilon = 0$; (b) $r = 3$, $a = 3$, $\upsilon = 1$; (c) $r = 5$, $a = 2$, $\upsilon = 1$; (d) $r = 7$, $a = 2$, $\upsilon = 2$.

We carry out the calculations for $a = 2$ to confirm the observation of Euler that we mentioned above, that everything depends on the remainder after dividing p by 8. Let $a = 2$; then it suffices to consider $r = 1, 3, 5, 7$, since in all other cases the arithmetic progression will contain no primes. As Figure 3 shows, 2 is a quadratic residue for $p = 8q + 1$ and $p = 8q + 7$, i.e., for $p = 8q \pm 1$.

Exercise. Show that -2 is a quadratic residue for $p = 8q + 1$, $p = 8q + 3$.

We treat $a = \pm 3$ analogously. Here is a table showing the values of υ:

a \ r	1	5	7	11
3	0	1	1	2
−3	0	1	2	3

Thus, 3 is a quadratic residue for $p = 12s \pm 1$ and a nonresidue for $p = 12s \pm 5$, and -3 is a quadratic residue for $p = 12s + 1$ and $p = 12s - 5$.

For $a = 2, 3$ you have of course noted yet another pattern: Primes whose remainders after dividing by $4a$ have the same absolute value are either both quadratic residues or both quadratic nonresidues. Euler did not fail to notice this, and he stated his hypothesis in a stronger form than we have.

FIGURE 4. $p = 11$, $q = 5$, $a = (p + q)/4 = 4$, $v(p) = 2$, $v(q) = 2$.

Supplement to Euler's Hypothesis. *Let p and q be primes with p + q = 4a. Then a is a quadratic residue for either both p and q or for neither.*

Proof. We carry out the construction in the proof of Euler's hypothesis for the intervals $(0, p/2)$ and $(0, q/2)$, with $a = (p + q)/4$. For convenience we arrange the intervals so that they are directed oppositely from 0, i.e., we reverse $(0, q/2)$ [Figure 4]. Let $v(p)$ and $v(q)$ be the number of integers in the respective odd-numbered intervals. It suffices to prove that $v(p) + v(q)$ is even. Let $v_j(p)$ and $v_j(q)$ be the number of integers in the respective jth intervals. It is easy to see that $v_j(p) + v_j(q) = 2$ for $j > 0$, which will imply the result we need.

Indeed, there are $2j$ integer points in the interval between the jth left and right points ($j > 0$), since as we have already noted, there are $2j$ integers in any interval of length $2j$ with noninteger endpoints.

The Law of Quadratic Reciprocity

In 1798, Legendre[5] pointed out a very convenient statement equivalent to Euler's hypothesis, the *law of quadratic reciprocity*. We introduce the following notation, known as the Legendre symbol:

$$\left(\frac{a}{p}\right) = \begin{cases} +1, \text{ if } a \neq 0 \text{ is a quadratic residue modulo } p \\ -1, \text{ if } a \text{ is a quadratic nonresidue.} \end{cases}$$

By Euler's criterion and the remark following it,

$$\left(\frac{a}{p}\right) = -a^{(p-1)/2} \equiv 0 \pmod{p}. \tag{7}$$

This immediately implies the multiplicative property of the Legendre symbol:

$$\left(\frac{ab}{p}\right) = \left(\frac{a}{p}\right)\left(\frac{b}{p}\right). \tag{8}$$

We also note that the definition of the Legendre symbol may be extended to all a not divisible by p so that (7) and (8) continue to hold, by setting

$$\left(\frac{a + p}{p}\right) = \left(\frac{a}{p}\right). \tag{9}$$

---

[5] Euler was apparently the first to conjecture the law of quadratic reciprocity, but both his and Legendre's proofs were incomplete. Gauss was the first to supply a correct proof. See Morris Kline, *Mathematical Thought from Ancient to Modern Times* (New York: Oxford Univ. Press, 1972), pp. 611–612, 814–815.—*Transl.*

The Law of Quadratic Reciprocity. *If p and q are odd primes, then*

$$\left(\frac{p}{q}\right)\left(\frac{q}{p}\right) = (-1)^{(p-1)/2 \cdot (q-1)/2}. \tag{10}$$

In other words, (p/q) and (q/p) have opposite signs if p = 4s + 3 and q = 4m + 3 and coincide otherwise.

This is called a "reciprocity" law because it establishes a reciprocity between p as a quadratic residue modulo q and q as a quadratic residue modulo p.

Proof. In all cases, either $p - q = 4a$ or $p + q = 4a$.

Case I. Suppose $p - q = 4a$, i.e., p and q have the same remainder after dividing by 4. Then $(p/q) = ((q + 4a/q) = (4a/q) = (a/q)$, by (9), (8), and the fact that $(4/q) = 1$ for all q. Moreover, $(q/p) = ((p - 4a)/p) = (-4a/p) = (-1/p)(a/p)$. By Euler's hypothesis, which we have already proved, $(a/p) = (a/q)$, i.e., $(p/q) = (q/p)$ when $(-1/p) = 1$ and $(p/q) = -(q/p)$ when $(-1/p) = -1$. It remains to recall that $(-1/p) = 1$ for $p = 4s + 1$ and $(-1/p) = -1$ for $p = 4s + 3$.

Case II. Suppose $p + q = 4a$, i.e., p and q have different remainders after dividing by 4. We have $(p/q) = ((4a - q)/q) = (a/q)$. Analogously, $(q/p) = (a/p)$. By the supplement to Euler's hypothesis, $(a/q) = (a/p)$, i.e., $(p/q) = (q/p)$. The proof is complete.

It is not hard to see that these arguments can be reversed to deduce Euler's hypothesis and its supplement from the quadratic reciprocity law (do it!). Also, formulas (8)–(10) give us a way to compute (p/q) that is substantially simpler than the combinatorial method described above. We illustrate it by an example: $(59/269) = (269/59) = ((59 \cdot 4 + 33)/59) = (3/59) \cdot (11/59) = -1$, since $(3/59) = (-59/3) = -(2/3) = 1$ while $(11/59) = -(59/11) = -(4/11) = -1$. It is easy to show that the computation of the Legendre symbol can always be reduced to the case where p or q equals 2.

Exercise. Compute $(37/557)$ and $(43/991)$.

In conclusion, we note that the problem of quadratic residues served as the starting point for a great and fruitful mathematical endeavor. Gauss' many attempts to obtain new proofs of the law of quadratic reciprocity were not primarily motivated by a desire to simplify its proof. The thought never left him that he had not really uncovered the deep patterns that gave rise to the law. These patterns were fully revealed only later, as part of the theory of algebraic numbers. Gauss spent a great deal of effort on generalizing the quadratic law to the cubic and biquadratic cases, obtaining remarkable results. This research has continued, and the study of various reciprocity laws remains one of the central problems of number theory to this day.

III. Royal Days

We have described in detail Gauss' first two great discoveries, made in Göttingen in a ten-day period a month before he turned nineteen. The second belongs entirely to arithmetic (number theory), while the first essentially depended on arithmetical considerations. Number theory was Gauss' first love.

The Favorite Science of the Greatest Mathematicians

This was one of the many names Gauss gave to arithmetic, i.e., number theory. At the time, arithmetic had already turned from a collection of isolated observations and assertions into a science.

Later, Gauss would write: "The really profound discoveries are due to more recent authors like those men of immortal glory P. de Fermat, L. Euler, L. Lagrange, A.M. Legendre (and a few others). They opened the door to what is penetrable in this divine science and enriched it with enormous wealth."[1]

One of the most surprising sides of the "Gauss phenomenon" is that in his earliest work he scarcely relied on the achievements of his predecessors, but instead quickly rediscovered what had been done in number

---

[1] *Disquisitiones Arithmeticae*, p. xviii.

The young Gauss, 1803.

theory during a century and a half of work by the best mathematicians.

Gauss used his stay in Göttingen to study the classic works, rethink them, and compare them with what he had himself discovered. In his view, the results of this activity should be summed up in a comprehensive work. Gauss set out to write such a book after returning to Braunschweig in 1798, after completing his university studies. It was to include his own results, which had remained unpublished if we do not count the newspaper notice, which incidentally promised: "This discovery is really only a corollary of a theory with greater content, which is not complete yet, but which will be published as soon as it is complete."[2] It took four years of strenuous work to realize this immense plan.

In 1801, Gauss' famous *Disquisitiones Arithmeticae* (*Investigations of Arithmetic*) appeared. This huge book (over 500 pages in large format) contains his fundamental results: the law of quadratic reciprocity, the problem of dividing a circle, and the question of representing integers in the form $am^2 + bmn + cn^2$ (in particular, as a sum of squares). The book was published with the support of the duke and was dedicated to him. As published, it consisted of seven parts. There was not enough money for an eighth, which was to be devoted to a generalization of the reciprocity law to powers greater than two, and in particular to the law of biquadratic reciprocity. Gauss did not find a complete proof of the latter until October 23, 1813, when he noted in his journal that this coincided with the birth of his son.

Klein wrote, "In the *Disquisitiones Arithmeticae* Gauss created modern number theory in its true sense and fixed its whole subsequent development. Our amazement at this achievement increases when we consider that Gauss created this whole world of thought purely out of himself and by himself, without any outside stimulus."[3]

Apart from *Disquisitiones Arithmeticae*, Gauss essentially discontinued his work in number theory. He only thought through and completed what he had conceived in those years. For example, he thought of six more proofs of the quadratic reciprocity law. *Disquisitiones* was far ahead of its time. Gauss had no serious mathematical contacts while writing it, and for a long time after the book appeared it was not intelligible to any of the German mathematicians. It was not brought to France, where he could have counted on the interest of Lagrange, Legendre, etc. The bookseller who was supposed to distribute it there went bankrupt, and a great many copies disappeared. As a result, Gauss' students later had to copy out parts of the book by hand. The situation in Germany began to change only in the 1840s, when Dirichlet studied *Disquisitiones* in depth and lectured on it. The book came to Kazan, to Bartels and his students, in 1807.

---

[2] Hall, *Gauss*, p. 24.
[3] *Development*, p. 24.

Disquisitiones turned out to have an enormous influence on the development of number theory and algebra. Starting from Gauss' work on dividing the circle, Galois successfully analyzed the solvability of equations in radicals. Reciprocity laws occupy a central position in algebraic number theory to this day.

The Helmstadt Dissertation

In Braunschweig, Gauss did not have the books he needed for his work on *Disquisitiones*. Thus he often went to neighboring Helmstadt, where there was a good library. Here in 1798, Gauss prepared a dissertation devoted to a proof of the Fundamental Theorem of Algebra, the assertion that every polynomial with complex (and, in particular, real) coefficients has a complex root. If we want to remain within the domain of real numbers, then the fundamental theorem takes the form: *every polynomial with real coefficients can be decomposed into the product of first- and second-degree polynomials*. Gauss critically reviewed all previous attempts at a proof and pursued an idea of D'Alembert with great care. Gauss' proof was not flawless, since a rigorous theory of continuity was still lacking. He later thought of three more proofs (the last one in 1848).

The Lemniscate and the Arithmetic-Geometric Mean

We will discuss one more direction in Gauss' work that he began in his youth.

In 1791, when Gauss was fourteen years old, he played the following game. He took two numbers a_0, b_0 and constructed their arithmetic mean $a_1 = (a_0 + b_0)/2$ and geometric mean $b_1 = \sqrt{a_0 b_0}$. Then he calculated the means of a_1 and b_1, i.e., $a_2 = (a_1 + b_1)/2$ and $b_2 = \sqrt{a_1 b_1}$, etc. Gauss computed both sequences to a large number of decimal places. Very soon he could no longer distinguish a_n from b_n—all the digits he calculated were the same. In other words, both sequences quickly approached the same limit $M(a_0, b_0)$, called the *arithmetic-geometric mean*.

In those years, Gauss tinkered a lot with a curve known as a *lemniscate* (or Bernoulli's lemniscate), the set of points for which the product of the distances from each of two fixed points O_1, O_2 (foci) is constant and equals $(\frac{1}{2} |O_1 O_2|)^2$. Gauss began a systematic study of the lemniscate in 1797. For a long time he tried to find its length, before guessing that it equals $2\pi|O_1 O_2|/M(\sqrt{2},2)$. We do not know how Gauss arrived at this, but we do know that it was on May 30, 1799, and that, not having a proof at first, he computed both quantities to eleven places (!). Gauss thought of a function for the lemniscate, analogous to the trigonometric functions for the circle. For example, for the lemniscate with foci $\sqrt{2}$ units apart, the lemniscate sine *sl t* is just the length of the chord corresponding to an arc of length *t*. Gauss spent the last years of the eighteenth century on constructing a

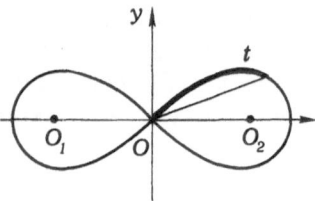

Lemniscate.

theory of lemniscate functions. He obtained addition and multiplication theorems, analogous to theorems for the trigonometric functions.

From lemniscate functions, Gauss turned to their generalization—elliptic functions. He understood that this was "a completely new domain of analysis." After 1800, he could no longer give elliptic functions the time needed to bring the theory to a state that was satisfactorily complete and rigorous. From the very beginning he avoided regular publication, hoping to publish everything at once, as he had done with his arithmetical work. But other matters never left him the necessary time.

In 1808, he wrote to his friend and student Heinrich Schumacher: "We now understand the circular and logarithmic functions as well as one times one, but the magnificent golden spring, guarding the secret of the higher functions, is still almost *terra incognita*. I worked very much on this before, and in time will present my own great work on it, which I alluded to in my *Disquisitiones Arithmeticae*. Come and be amazed at the extraordinary richness of the new and most interesting truths and relationships possessed by these functions."

Gauss believed that he did not have to hurry to publish his results. It went on like this for thirty years, and then in 1827 two young mathematicians, Abel and Jacobi, published much of what he had obtained.

"Jacobi's results represent a part of my own great work, which I intended to publish at some time. It will be an exhaustive work on this subject, if only heaven will be pleased to prolong my life and give me strength and peace of mind" (from a letter to Schumacher).

"Mr. Abel anticipated many of my ideas and made my problem easier by about a third, setting forth results with great rigor and elegance. Abel followed the same path that I did in 1798, so there is nothing improbable in the fact that we obtained so similar results. To my surprise, this resemblance even extends to form, and here and there to notation, therefore many of his formulas seem copied from mine. But so that no one should misunderstand me, I must add that I do not recall a single instance when I talked about this research with any outsider" (from a letter to Bessel).

Finally, in a letter to Crelle: "Since Abel has demonstrated such insight and such elegance in the problems in his account, I feel that I may completely refrain from publishing the results I obtained" (May 1828).

We should note that Gauss' remark in *Disquisitiones* that the theory of dividing a circle could be extended to the lemniscate turned out to have a great influence on Abel.

With the beginning of the new century, Gauss' scientific interests shifted away from pure mathematics. He returned to it periodically and each time obtained new results worthy of his genius. In 1812, he published a work on the hypergeometric function. (This function depends on three parameters, and by giving them specific values we can obtain most of the functions occurring in mathematical physics). Gauss' contribution to the geometric interpretation of complex numbers is widely known. We will discuss his work in geometry below. But mathematics would no longer be the main work of his life. A characteristic external sign is that in 1801 he stopped making regular entries in his journal, although there are notes until 1814. We rarely realize how short Gauss' "mathematical century" was —less than ten years—and a great part of this time was devoted to work that was unknown to his contemporaries (elliptic functions).

Asteroids

We will now relate Gauss' new inclinations. Many biographers have argued about the reasons why Gauss began to study astronomy. We must first keep in mind that, beginning with Kepler, Galileo, and Newton, astronomy was the most striking area in which to apply mathematics. This tradition was continued in the work of Euler, D'Alembert, Clairaut, Lagrange, and Laplace. In predicting and explaining celestial phenomena, mathematicians felt as if they were admitted to the mysteries of the universe. Gauss, with his early interest in concrete calculations, could of course not help but try his strength in this traditional arena.

There were also prosaic reasons. Gauss earned a meager living as a *privat-dozent* in Braunschweig, receiving 6 thalers a month. A pension of 400 thalers from his patron duke did not improve his situation enough to support a family, and he was contemplating marriage. It was not simple to obtain a chair in mathematics somewhere, and Gauss was not very attracted to teaching. The broadening net of observatories made a career as an astronomer more accessible.

Gauss had begun to be interested in astronomy while in Göttingen. He carried out some observations in Braunschweig, and used part of his pension from the duke to buy a sextant. He searched for a worthy computational problem, solving minor ones in the meantime. For example, he published a simple method for computing the times of Easter and other cyclical holidays, in place of the extremely confusing techniques that had been used earlier. The idea for a real problem arose in 1801, under the following circumstances.

On January 1, 1801, the astronomer Piazzi, who was creating a star catalog, discovered an unknown star of the eighth magnitude. He observed it for forty days and then asked the leading astronomers to continue the observations. For various reasons, his request was not met. In June the information reached Zach, who published the only journal in astronomy at the time. Zach conjectured that this was a "long supposed, now probably discovered, new major planet of our solar system between Mars and Jupiter." Zach's conjecture seemed plausible, and an urgent search for the "missing" planet was required. But this demanded the calculation of its trajectory. Determining an elliptical trajectory from the 9° arc that Piazzi had found was beyond the limits of the computational ability of the astronomers. In September 1801, Gauss dropped everything else and undertook to compute the orbit. In November the calculations were complete. They were published in the December issue of Zach's journal, and on New Year's Eve, exactly a year after Piazzi's sighting, the famous German astronomer Heinrich Olbers, using the trajectory calculated by Gauss, found the asteroid now known as Ceres. It was a genuine sensation!

On March 25, 1802, Olbers discovered another asteroid, Pallas. Gauss quickly computed its orbit, showing that it lay between Mars and Jupiter. The astronomers unhesitatingly accepted the validity of Gauss' computational methods.

Recognition came to Gauss. including election as a corresponding member of the St. Petersburg Academy of Sciences. He was soon invited to take the position of director of the observatory there. Gauss wrote that it was flattering to be invited to the city where Euler had worked, and that he was seriously considering the move. In his letters, Gauss wrote that the weather was often bad in St. Petersburg, so that he would not be too busy with observation and would have time for study. The thousand rubles he would receive were worth more than the 400 thalers he now had, but it was more expensive to live there.

At the same time, Olbers began efforts to keep Gauss for Germany. In 1802 he proposed that the curators of the University of Göttingen offer Gauss the position of director of the newly organized observatory there. Olbers wrote that Gauss "has a positive aversion to a chair in mathematics." It was agreed on, but Gauss did not move until the end of 1807. During that time he married ("life seems to me springlike, always with bright new colors."). In 1806 the duke, for whom Gauss evidently had a sincere attachment, died from a wound. Now there was nothing to keep him in Braunschweig.

Gauss' life in Göttingen did not proceed sweetly. In 1809, his wife died after giving birth to a son, and then the child died, too. In addition, Napoleon placed a heavy tax on Göttingen, and Gauss himself had to pay the unbearable amount of 2000 francs. Olbers tried to pay the money for him, and so did Laplace, who was right in Paris. Both times Gauss proudly refused. Another benefactor was found, this time anonymous, and there was no one to whom Gauss could return the money. It was learned much later that this was the Elector of Mainz, a friend of Goethe. "Death is dearer to me than such a life," wrote Gauss, between notes on the theory of elliptic functions. Those around him did not value his work and considered him an eccentric, at the least. Olbers calmed Gauss, saying he shouldn't count on people's understanding: "You must take pity on them and serve them."

The year 1809 also saw the appearance of the famous *Theoria Motus Corporum Coelestium* (*Theory of Motion of the Heavenly Bodies, Moving About the Sun in Conic Sections*),[4] which Gauss had completed in 1807. The delay occurred partly from the publisher's fear that there would be no demand for the book in German, and for patriotic reasons.Gauss refused to publish it in French. They compromised by publishing it in Latin. This was Gauss' only book on astronomy, although he also published a few articles.

Gauss set out his methods for computing orbits. To persuade the reader of the power of his technique, he repeated the calculation of the orbit of the comet of 1769, which Euler had found at the time after three days of intensive computation (losing sight of it afterwards, according to some sources). Gauss required one hour. The book contained the method of least squares, which remains to this day one of the most widely used methods for working with observational data. Gauss indicated that he had known of this method since 1794, and had used it systematically since 1802. (Legendre had published the method of least squares two years before the appearance of Gauss' *Theoria Motus*.)

Gauss received many honors in 1810: a prize from the Academy of Sciences of Paris and the Gold Medal of the Royal Society of London, as well as election to several academies.

In 1804, the Paris Academy took the theory of perturbations of Pallas as the theme for its grand prize (a gold medal weighing 1 kilogram). The deadline was twice extended (to 1816) in the hope that Gauss would submit an entry. Gauss was assisted in his calculations by his student Nicolai ("a young man, tireless in carrying out calculations"), but they were still not carried through to the end. Gauss stopped working on them, falling into a deep depression.

Gauss continued his regular astronomical activities almost up to his death. The famous comet of 1812 (which "presaged" the fire of Moscow!) was observed everywhere, thanks to his calculations. On August 28, 1851, he observed a solar eclipse. He had many astronomy students (Schumacher, Gerling, Nicolai, Struve). The excellent German geometers Möbius and Staudt studied not geometry but astronomy with him. He conducted an active correspondence with many astronomers, regularly read articles and books on astronomy, and published reviews. We also

---

[4] There is an 1857 English translation of this book by Charles Henry Davis, recently reprinted (New York: Dover, 1963).—*Transl.*

learn much about his mathematical activities from his letters to astronomers. How different the image of Gauss the astronomer is from the inaccessible recluse that he seems to mathematicians!

Geodesy

By 1820, the center of Gauss' practical interests had shifted to geodesy. As early as the beginning of the century, he had tried to use measurements of a meridian arc, made by French geodesists to establish a standard of length (the meter), to find the true amount of the earth's flattening.[5] But the arc turned out to be too small. Gauss dreamt of measuring a sufficiently large arc, but was only able to begin in 1820. Although the measurement stretched out over two decades, Gauss could not fully realize his idea. His research on the treatment of the results of measurements, carried out in connection with geodesy, were of great importance (his fundamental publications on the method of least squares belong to this time), as well as various geometric results related to the need for making measurements on the surface of an ellipsoid.

In the twenties, the question arose of Gauss moving to Berlin, where he would become the head of an institute. The most promising young mathematicians were to be invited there, above all Jacobi and Abel. The negotiations were drawn out over four years. The disagreement was over whether Gauss would deliver lectures and how much he would be paid— 1200 or 2000 thalers a year. The negotiations were unsuccessful, but not completely: In Göttingen, Gauss was paid the salary he wanted in Berlin.

The Inner Geometry of a Surface

We are obliged to geodesy for the fact that, for a comparatively short time, mathematics again became one of Gauss' main activities. In 1816 he thought of generalizing the basic problem of cartography, mapping one surface onto another "so that the image is similar to the original in the smallest parts." Gauss advised Schumacher to select this problem in competing for the prize of the Copenhagen Scientific Society. The competition was declared in 1822. In the same year Gauss submitted his memoir, in which he introduced parameters[6] allowing a complete solution of the problem, special cases of which had been studied by Euler and Lagrange (mapping a sphere or surface of revolution onto a plane). Gauss described in detail the conclusions of his theory for many concrete cases, some of which arise in geodesy problems.

In 1828, Gauss' fundamental geometry memoir, *Disquisitiones Generales circa Superficies Curvas* (*General Investigations of Curved Surfaces*),

---

[5] i.e., the extent to which its shape is not spherical—*Transl.*
[6] curvilinear coordinates—*Transl.*

appeared.[7] It is devoted to the inner geometry of a surface, i.e., to what is associated with the very structure of the surface and not to its position in space.

Figuratively speaking, the inner geometry of a surface is what we can learn about its geometry "without leaving it." On the surface we can measure length, by stretching a string so that it lies completely on the surface. The resulting curve is called a geodesic (analogous to a line on a plane). We can measure angles between geodesics and study geodesic triangles and polygons. If we deform the surface (thinking of it as an unstretchable and untearable film), then distances between points will be preserved, geodesics will remain geodesics, etc.

It turns out that, "without leaving the surface," we can tell whether or not it is curved. A deformation can never turn a "really" curved surface into a plane. Gauss proposed a numerical measure for the degree of curvature of a surface.

Consider a neighborhood of area ε near a point A of a surface. At each point of this neighborhood we take a normal (a vector perpendicular to the surface) of length one. All normals of a plane will be parallel, but for a curved surface they will diverge. We displace the normals so that their origins are at a single point. Then their endpoints form some figure on the unit sphere. Let $\varphi(\varepsilon)$ be the area of this figure. Then

$$k(A) = \lim_{\varepsilon \to 0} \frac{\varphi(\varepsilon)}{\varepsilon}$$

measures the curvature of the surface at A. It turns out that $k(A)$ is the same for all deformations. In order for a piece of the surface to turn into a plane, $k(A)$ must be zero at all points A of the piece. This measure of curvature is associated with the sum of the angles of a geodesic triangle.

Gauss was interested in surfaces of constant curvature. A sphere is a surface of constant positive curvature (at each of its points $k(A) = 1/R$, where R is the radius). In his notes, Gauss mentions a surface of revolution with constant negative curvature. Later it was a called a *pseudosphere*, and Beltrami discovered that its inner geometry is a Lobachevski non-Euclidean geometry.

Non-Euclidean Geometry

According to some information, Gauss had even been interested in the parallel postulate in 1792, while in Braunschweig. In Göttingen he often discussed this problem with Bolyai Farkas,[8] a Hungarian student. We know from a 1799 letter to Bolyai how clearly Gauss understood that there

---

[7] There is an English translation by Adam Hiltebeitel and James Morehead (Hewlett, NY: Raven Press, 1966).—*Transl.*
[8] sometimes known as Wolfgang Bolyai—*Transl.*

are many assertions which, if we accept, allow us to prove the fifth postulate. "I have certainly achieved results which most people would look upon as a proof...." And, "...the way in which I have proceeded does not lead to the desired goal but instead to doubting the validity of geometry." It is only one step from here to understanding the possibility of non-Euclidean geometry, but that step still had not been taken. This sentence is often erroneously taken as evidence that Gauss had arrived at non-Euclidean geometry as early as 1799.

Gauss' words that he is not able to devote enough time to this problem deserve notice. It is typical that there is no mention in his journal of the problem of parallel lines. It was evidently never at the center of Gauss' attention. In 1804, he rejected Bolyai's attempts to prove the parallel postulate. His letter ends "However I still hope that at some time, and before my end, these submerged rocks will allow us to pass over them." These words seem to indicate a hope that a proof would be found.

Here is more testimony: "In the theory of parallel lines we are now no further than Euclid was. This is the *partie honteuse* [shameful part] of mathematics, which sooner or later must get a very different form" (1813). "We have not advanced beyond the place where Euclid was 2000 years ago" (1816). But in that very year of 1816 he speaks of "the gap which cannot be filled," and in 1817 we read in a letter to Olbers: "I am coming more and more to the conviction that the necessity of our geometry cannot be proved, at least not *by* human intelligence and not *for* human intelligence. Perhaps we shall arrive in another existence at other insights into the essence of space, which are now unattainable for us. Until then one would have to rank geometry not with arithmetic, which stands a priori, but approximately with mechanics."[9]

At about the same time, F.C. Schweikart, a jurist from Königsberg, arrived at the notion that the fifth postulate was impossible to prove. He proposed that an "astral geometry," in which the parallel postulate does not hold, exists alongside Euclidean geometry. Gauss' student C.L. Gerling, who was working in Königsberg, wrote to his teacher about Schweikart's idea and attached a note from him. In reply, Gauss wrote: "Almost everything is copied from my soul." Schweikart's activities were continued by his nephew Taurinus, with whom Gauss exchanged some letters, beginning in 1824.

In his letters Gauss stressed that his statements were of an especially partial nature and must in no case be made public. He did not believe these ideas could be grasped, and feared the interest of hordes of dilettantes. Gauss had spent more than a few difficult years and greatly valued the opportunity to work quietly. He warned Gerling, who had planned only to

---

[9] This translation is taken from G. Waldo Dunnington, *Carl Friedrich Gauss*: *Titan of Science* (New York: Exposition Press, 1955), p. 180.

mention that the parallel postulate could turn out to be false, "but the wasps, whose nest you stir up, will fly around your head." He gradually reached a decision to write down his results but not publish them: "Probably, I will not be able to work out my space research on this question soon enough so that they can be published. Perhaps it will not happen during my lifetime, since I fear the Bœotians'[10] cries if I were to express my opinion fully" (from an 1829 letter to Bessel). In May of 1831, Gauss began to make systematic notes: "It has now been several weeks since I began to set out in writing several results of my own thinking on this subject, in part already forty years old but that I never wrote down, as a result of which I had to begin the whole work over again three or four times in my head. I did not want, however, for this to buried along with me" (letter to Schumacher).

But in 1832 he received from Bolyai Farkas a short essay by his son János, *Appendix Scientiam Spatii Absolute* (*The Science of Absolute Space*) (so called because it was published as an appendix to a large book written by the father). "My son values your opinion more than the opinion of all Europe." The content of the book startled Gauss: a complete and systematic construction of non-Euclidean geometry. These were not the fragmentary remarks and conjectures of Schweikart-Taurinus. Gauss himself had intended to produce such an account in the near future. He wrote Gerling: "I found all my own ideas and results, developed with great elegance, although because of the conciseness of the account, in a form that is accessible with difficulty to someone to whom this area is foreign ....I believe that this young geometer Bolyai is a genius of the first order." And to the father, he wrote "...all of the paper's contents, and the way your son has attacked the matter, coincide almost completely with my own reflections which I partly carried out thirty to thirty-five years ago. In fact I am extremely surprised by it. My intention was to leave my own work, of which at the present time only a small part is written down, unpublished during all of my lifetime....On the other hand I had intended to write it all down little by little, so that it at least would not disappear with me. I am thus quite surprised that I can spare myself these efforts, and it makes me very happy that it is the son of one of my old friends who has come ahead of me in such a remarkable manner."[11] Bolyai János received no public appreciation or support from Gauss whatsoever. Gauss evidently interrupted his systematic notes on non-Euclidean geometry at that time, although sporadic notes from the 1840s remain.

In 1841 Gauss became acquainted with the German edition of Lobachevsky's works (Lobachevsky's first publications date from 1829).

---

[10] According to legend, the inhabitants of Bœotia were famous in ancient Greece for their stupidity.
[11] This translation is taken from Hall, *Gauss*, p. 114.

True to form, Gauss was interested in the author's other publications, but commented about him only in letters to his close correspondents. Incidentally, on Gauss' proposal, Lobachevsky was elected in 1842 as a corresponding member of the Göttingen Royal Society, "as one of the most excellent mathematicians of the Russian people." Gauss personally notified Lobachevsky of his election. However, non-Euclidean geometry was not mentioned in either Gauss' presentation or the diploma given Lobachevsky.

We know of Gauss' work on non-Euclidean geometry only from the posthumous publication of his archives. Gauss thus guaranteed that he could work quietly by refusing to publicize his great discovery, resulting in arguments that continue to this day about the appropriateness of his position.

We should note that Gauss was interested in more than the purely logical question of the provability of the parallel postulate. He was interested in the place of geometry in the natural sciences and the truth of the geometry of our physical world (see his above statement of 1817). He discussed the possibility of an astronomical verification, speaking with interest of Lobachevsky's ideas in this regard. In his work on geodesy, Gauss even measured the sum of the angles of the triangle formed by the German mountains of Hohenhagen, Brocken, and Inselsberg. It differed from 2π by no more than $0.2'$.

Electrodynamics and Terrestrial Magnetism

By the end of the 20s Gauss, now on the far side of fifty, began to look for areas of science that were new for him. This is seen in two publications of 1829 and 1830. The former set out his thoughts on general principles of mechanics (his "Principle of Least Constraint"), while the latter was devoted to the study of capillary phenomena. Gauss decided to study physics, but had not yet defined his specific interests. In 1831 he tried to study crystallography. This was a very difficult year in his life: His second wife died, and he began to suffer from terrible insomnia. In the same year, the twenty-seven-year-old physicist Wilhelm Weber was invited to Göttingen, at Gauss' initiative. Gauss had met him in 1828, at Humboldt's house. In 1831 Gauss was fifty-four. His reserve had become legendary, but all the same he found a scientific partner in Weber as he had never had before.

"The inner difference between the two men was also expressed quite clearly in their outer appearance. Gauss had a powerful, stocky physique, a true Lower Saxon, laconic and not easily accessible. This was in strong contrast to the small, delicate, agile Weber, whose friendly, loquacious nature betrayed at once the true Saxon, though he was actually born in Wittenberg, in the land of the "double Saxon" [Doppelsachsen]. In the Gauss-Weber Monument in Göttingen this contrast has been minimized

for artistic reasons, and even their ages appear closer than they were"
(Klein)[12].

Gauss' and Weber's interests lay in the area of electrodynamics and
terrestrial magnetism. Their work had not only theoretical but also
practical results. In 1833 they invented an electromagnetic telegraph (this
event is recorded on their common memorial). The first telegraph connected the observatory and the physics institute but, for financial reasons, its creators did not succeed in developing it further.

In the course of his research, Gauss arrived at the conclusion that an
absolute system of physical units was needed for work in magnetism. He
began with a number of independent quantities and expressed the remaining quantities in terms of them.

The study of terrestrial magnetism relied both on observations at the
magnetic observatory at Göttingen, and on materials that had been

---

[12] Development, pp. 17–18.

collected in various countries by the "Society for the Observation of Terrestrial Magnetism," created by Humboldt after his return from South America. At the same time, Gauss created one of the most important chapters in mathematical physics—potential theory.

Gauss' and Weber's joint work was interrupted in 1843, when Weber and six other professors were expelled from Göttingen after signing a letter to the king that cited breaches in the recent constitution (Gauss had not signed it). Weber returned to Göttingen only in 1849, when Gauss was already seventy-two years old.

We conclude our story of Gauss with Klein's words: "To me Gauss is like the highest peak amidst our Bavarian mountains as it appears to a spectator from the north. From the east the gradually ascending foothills culminate in the one gigantic colossus, which falls away steeply into the lowlands of a new formation, into which its spurs project for many miles and in which the water that streams from it begets new life."[13]

Appendix: Construction Problems Leading to Cubic Equations

In *Disquisitiones Arithmeticae*, Gauss states without proof that it is impossible to construct, with straightedge and compass, regular n-gons for primes n that are not Fermat primes, and in particular, a regular 7-gon. This negative result must have surprised his contemporaries no less than the possibility of constructing a regular 17-gon. After all, $n = 7$ is the first value of n for which no construction was found, despite many attempts. The Greek geometers undoubtedly suspected there was something troublesome about this problem and it was not without cause, let us say, that Archimedes proposed a method for constructing a regular 7-gon using conic sections. However, the question of proving that the construction was impossible evidently did not even arise.

One must say that proofs of negative assertions have always played a fundamental role in the history of mathematics. An impossibility proof requires that every conceivable method of solution, construction, or proof somehow be considered, while it suffices to give just one specific method for a positive solution.

Impossibility proofs in mathematics had a famous beginning, when the Pythagoreans (sixth century B.C.), in trying to reduce all of mathematics to integers, buried this area with their own hands: It turned out that there exists no fraction whose square equals 2. Another way to say this is that the diagonal and sides of a square are incommensurable. Thus integers and their ratios are insufficient to describe a very simple situation. This dis-

---

[13] *Development*, p. 57.

covery surprised the greatest thinkers of ancient Greece. Legend states that the gods punished the Pythagorean who communicated this fact (he died in a shipwreck). Plato (429–348 B.C.) tells how the existence of irrational quantities astonished him. Plato once ran into a "practical" problem that caused him to rethink the possibilities of geometry.

"In his work entitled Platonicus, *Eratosthenes relates* that, when God announced to the Delians through an oracle that, in order to be liberated from the pest, they would have to make an altar, twice as great as the existing one, the architects were much embarassed in trying to find out how a solid could be made twice as great as another one. They went to consult Plato, who told them that the god had not given the oracle because he needed a doubled altar, but that it had been declared to censure the Greeks for their indifference to mathematics and their lack of respect for geometry" (Theon of Smyrna.)[14] Don't say that Plato didn't make use of the right moment for scientific propaganda! According to Eutocius, an analogous problem (doubling the volume of Glaucus' tombstone) figured in one version of the legend of Minos.

We are talking about finding the sides of a cube with twice the volume of a given cube, i.e., of constructing the roots of the equation $x^3 = 2$. Plato sent the Delians to Eudoxus and Helicon. Menæchmus, Archytas, and Eudoxus proposed various solutions, but none found a construction with straightedge and compass. Eratosthenes, who built a mechanical device for solving the problem, later called his predecessors' solutions too complicated, in a poem carved on a marble slab in Ptolemy's temple in Alexandria: "Do not thou seek to do the difficult business of Archytas' cylinders, or to cut the cone in the triads of Menæchmus, or to compass such a curved form of lines as is described by the god-fearing Eudoxus."[15] Menæchmus noted that the problem is equivalent to finding two mean proportionals, i.e., for given a and b finding x and y that satisfy $a:x = x:y = y:b$. The latter problem was solved using conic sections. We know nothing of Eudoxus' "curved lines." As far as a mechanical solution is concerned, Eratosthenes was not the first. According to Plutarch, "Plato himself censured those in the circles of Eudoxus, Archytas, and Menæchmus, who wanted to reduce the duplication of the cube to mechanical constructions, because in this way they undertook to produce two mean proportionals by a nontheoretical method; for in this manner the good in geometry is destroyed and brought to nought, because geometry reverts to observation instead of raising itself above this and adhering to the eternal, immaterial images, in which the immanent God is the eternal God."[16]

---

[14] As translated in Bartel L. van der Waerden, *Science Awakening I*, trans. Arnold Dresden (New York: Oxford University Press), p. 161.
[15] Ibid.
[16] Ibid., p. 163.

Incidentally, Eutocius ascribes to Plato himself (evidently erroneously) a mechanical solution to the Delian problem, using a carpenter's square with grooves and adjustable rulers. Plato, with his aversion to "material things, which require extended operations with unworthy handicrafts" (Plutarch)[17], is often contrasted to Archimedes (287–212 B.C.), who was glorified for his many inventions, in particular for the machines used in the defense of Syracuse. The same Plutarch also claimed that Archimedes only yielded to the persuasion of King Hiero "to direct his art somewhat away from the abstract...[and to occupy] himself in some tangible manner with the demands of reality," although he believed that the practical was "lowbrow and ignoble, and he only gave his efforts to matters which, in their beauty and their excellence, remain entirely outside the realm of necessity."[18]

Along with the Delian problem, Greek geometry left several other problems for which a construction with straightedge and compass was not found: trisecting an angle (dividing it into three equal parts), squaring the circle, and constructing a regular n-gon, in particular a 7-gon and a 9-gon. The Greeks, and even more so the Arab mathematicians, were aware of the connection between these problems and cubic equations.

The problem of the regular 7-gon reduces to the equation $z^6 + z^5 + z^4 + z^3 + z^2 + z + 1 = 0$ (see p. 120), or

$$\left(z^3 + \frac{1}{z^3}\right) + \left(z^2 + \frac{1}{z^2}\right) + \left(z + \frac{1}{z}\right) + 1 = 0.$$

Passing to the variable $x = z + 1/z$, we obtain the equation $x^3 + x^2 - 2x - 1 = 0$.

We will show that the roots of the equations for doubling the cube and for the 7-gon cannot be quadratic irrationals, which will imply the impossibility of constructing them with straightedge and compass. We will prove a result that applies to a rather general situation.

Theorem. *If a cubic equation $a_3 x^3 + a_2 x^2 + a_1 x + a_0 = 0$ with integer coefficients has a quadratic irrational root, then it also has a rational root.*

Proof. Let x_1 be a quadratic irrational root. Then x_1 can be obtained from integers by arithmetic operations and extracting square roots; let us analyze this construction. We first take the square roots of certain rational numbers, $\sqrt{A_1}, \sqrt{A_2}, \ldots, \sqrt{A_a}$, and then of certain numbers obtained from rationals and the $\sqrt{A_i}$, say $\sqrt{B_1}, \sqrt{B_2}, \ldots, \sqrt{B_b}$, etc. At each step, we take roots of numbers that are expressed arithmetically in terms of those obtained previously. We thus obtain "stages" of quadratic irrationals. Let

---

[17] Ibid.
[18] Ibid., p. 209.

$\sqrt{N}$ be one of the numbers obtained at the last step before forming x_1. We fix our attention on how $\sqrt{N}$ enters into x_1. It turns out that we may assume that x_1 has the form $\alpha + \beta \sqrt{N}$, where α, β are quadratic irrationals into which $\sqrt{N}$ does not enter. It suffices to note that arithmetic operations on expressions of the form $\alpha + \beta \sqrt{N}$ lead to the same type of expression. This is obvious for addition and subtraction and can be directly verified for multiplication. For division we must eliminate $\sqrt{N}$ from the denominator:

$$\frac{\alpha + \beta\sqrt{N}}{\gamma + \delta\sqrt{N}} = \frac{(\alpha + \beta\sqrt{N})(\gamma - \delta\sqrt{N})}{\gamma^2 - \delta^2 N}.$$

If we now substitute $x_1 = \alpha + \beta \sqrt{N}$ into the equation and carry out the operations, we obtain a relation of the form $P + Q\sqrt{N} = 0$, where P and Q are polynomials in α, β, a_i. If $Q \neq 0$ then $\sqrt{N} = P/Q$, and by substituting this into the expression for x_1 we can represent x_1 without $\sqrt{N}$. If $Q = 0$, then it can be verified that $x_2 = \alpha - \beta\sqrt{N}$ is also a root, and since $-a_2/a_3 = x_1 + x_2 + x_3$ is the sum of the roots (Vieta's theorem), we obtain $x_3 = (-a_2/a_3) - 2\alpha$. Thus we again have a root that is a quadratic irrational, expressible in terms of $\sqrt{A_i}, \sqrt{B_j}, \ldots$, as is x_1, but without $\sqrt{N}$. Continuing this process further, we eliminate all radicals in the expression for a root of the equation by stages, beginning with the last stage. After this we obtain a rational root, and the proof is complete.

It now remains to verify that the equations we are interested in have no rational roots. Suppose an equation has leading coefficient $a_3 = 1$. Then each rational root is an integer. It suffices to substitute $x = p/q$ (p, q relatively prime) into the equation, multiply both sides by q^3, and see that p^3 and thus p are divisible by q, i.e., $q = 1$. Furthermore, if α is a root, then $x^3 + a_2 x^2 + a_1 x + a_0 = (x - a)(x^2 + mx + n)$, where $a_2 = -\alpha + m$, $a_1 = -\alpha m + n$, $a_0 = -\alpha n$, i.e., $m = a_2 + \alpha$, $n = a_1 + a_2 \alpha + \alpha^2$. This means that if a_i and α are integers, then m and n are integers and α must divide a_0. So with $a_3 = 1$, the search for rational roots of the equation reduces to sorting out a finite number of possibilities, the divisors of the constant term. It is now easy to verify that the equations we are considering have no integer roots, so they have no roots that are quadratic irrationals.